Advanced Satellite Technologies

ADVANCED SATELLITE TECHNOLOGIES

Edited by:

Dr. Kailash Malode and Dr. Mahesh Deshmukh

ARCLER PRESS

www.arclerpress.com

Advanced Satellite Technologies

Dr. Kailash Malode and Dr. Mahesh Deshmukh

Arcler Press
224 Shoreacres Road
Burlington, ON L7L 2H2
Canada
www.arclerpress.com
Email: orders@arclereducation.com

ISBN: 978-1-77469-833-4 (Hardcover)

ABOUT THE EDITORS

Dr. Kailash Malode (1980) is presently serving as Assistant Professor, Department of Soil Science And Agriculture Chemistry, college of Agriculture, Parul University, Vadodara, Gujarat. He obtained his B. Sc.(Ag.) in 2006 from Marathwada Agriculture University, Parbhani. M. Sc. (Ag.)2009 in Dr. Panjabrao Deshmukh Krishi Vidyapeeth, Akola and Ph. D. Soil Science And Agriculture Chemistry in 2014 from Vasantrao Naik Marathwada Krishi Vidhyapeeth, Parbhani. He started his career as Assistant professor in 2014. His field of specialization is soil fertility, soil carbon sequestration and organic carbon management. He has edited 06 books at international level, published 19 research papers in National and international journals of reputed. He has participated in many National, State Seminars and symposiums. He has written 02 practical manuals for the undergraduate course for the benefit of students, besides several radio talks and on farmers on a farmer training programme. Dr. Kailash has also written 26 popular articles in Shati Bharti, Shati pragati and other reputed magazines and 22 audio talks. He had the Best writer award in 2015 for Krishi dooth popular Magazines of Maharashtra, Young Scientists Award 2021 by Astha foundation and Birsa Agriculture University, Jharkhand. Dr. Kailash got the Best scientist award in soil science 2023 by Rajmata Vijayaraje Sindhya Krishi Vishva Vidhyalaya, Gwalior, and also best poster presentation of same.

Dr. Mahesh Sharadrao Deshmukh presently working as Assistant professor at College of Agriculture, VNMKV Parbhani. He did his B.Sc.(Agri) in 1998 from VNMKV Parbhani, Post Graduation in Dr. PDKV, Akola during 2000 and PhD from SRTMU, Nanded in 2008. His area of specialization is in soil fertility mapping, INM and organic farming. He has published 03 books, 29 Research Paper,22 popular articles. Dr. Mahesh also received Best Poster Presentation Award in National Seminar on State Agriculture University.

TABLE OF CONTENTS

List of Figures ... *xi*

List of Tables ... *xv*

List of Abbreviations ... *xvii*

Preface ... *xxi*

Chapter 1 Introduction to Satellites and Their Applications ... 1

1.1. Introduction ... 2

1.2. What is the Satellite? ... 3

1.3. The Era of Sounding Rockets and Hot Air Balloons ... 6

1.4. Launch of the Early Artificial Satellites ... 7

1.5. Satellites For Meteorology, Scientific Exploration, and Communications: Early Developments ... 10

1.6. Non-Geosynchronous Communication Satellites: Relay and Telstar Programs ... 11

1.7. Development of the Geosynchronous Communication Satellites ... 13

1.8. Global Communication Satellite Systems ... 15

1.9. Local Communication Satellite Systems ... 18

1.10. Satellites for Various Other Applications Also Made Quick Progress ... 21

1.11. Evolution of the Launch Vehicles ... 26

1.12. Future Trends ... 32

References ... 35

Chapter 2 Classification of Key Satellite Systems ... 43

2.1. Introduction ... 44

2.2. Geosynchronous Earth Orbit Satellites ... 49

2.3. Medium Earth Orbit (Meo) Satellites ... 51

2.4. Low Earth Orbit (LEO) Satellites ... 53

2.5. Highly Elliptical Orbit (HEO), Basics ... 56

2.6. Polar Orbit and Sun-Synchronous Orbit (SSO) 58
2.7. Transfer Orbits and Geostationary Transfer Orbit (GTO) 59
2.8. Lagrange Points 60
2.9. Narrowband Leo, Meo, and Geo Systems 60
References 64

Chapter 3 Introduction to Low Earth Orbital (LEO) Satellites 71
3.1. Introduction 72
3.2. Leos Economic Viability 74
3.3. Leos Research Issues 76
3.4. Summary and Discussion 88
References 90

Chapter 4 Medium Earth Orbit Satellites 99
4.1. Introduction 100
4.2. Satellite Constellation Architecture Design Tradeoffs 100
4.3. Regional Layout and Gateways 103
4.4. Low Latency Advantages For Interactive Applications 103
4.5. Impact of High Latency on File Transfer and Web-Based C^2 Applications Over Tcp/Ip 105
4.6. High Data Rate Advantages 107
4.7. Implications for Ground Terminal Equipment 107
References 109

Chapter 5 Geostationary Satellites and Their Constellations 113
5.1. Introduction 114
5.2. Development of Geostationary Satellites 116
5.3. Satellite Constellations: Many Architectures and Many Problems 119
References 121

Chapter 6 An Overview of Cube-Satellite Propulsion Technologies and Trends 125
6.1. Introduction 126
6.2. Cold Gas Propulsion (CGP) Systems 128
6.3. Liquid Propulsion (LP) Systems 131
6.4. Solid Rocket Propulsion (SRP) Systems 135
6.5. Resistojets 138
6.6. Radio-Frequency Ion Thruster (RIT) 141

6.7. Hall Effect Propulsion/Hall Thrusters 144
References 147

Chapter 7 Small Satellites Missions 155
7.1. Introduction 156
7.2. Classification 160
7.3. Ten-Koh Project 172
References 179

Chapter 8 Future Trends in Satellite Communication Systems 187
8.1. Introduction 188
8.2. Current Trends in Satellite Technology 189
8.3. The Path Forward 190
8.4. Satellite Orbital Configurations and Improved Spacecraft Orientation and Pointing 202
8.5. New Ground User Systems 204
8.6. Integrated Satellite and Terrestrial Services and New Market Demand 205
8.7. Tracking, Telemetry, Monitoring, Command, and Autonomous Operations 207
8.8. Future Trends for Markets and Regulatory Systems 208
8.9. New Initiatives in Space Communications 210
8.10. Keys to the Future of Satellite Communications 211
References 216

Index 227

LIST OF FIGURES

Figure 1.1. A communication satellite
Figure 1.2. An earth observation satellite
Figure 1.3. Weather forecasting satellite
Figure 1.4. United States military satellite
Figure 1.5. Scientific satellite
Figure 1.6. 1st A-4 rocket to be taken off
Figure 1.7. Sputnik-1
Figure 1.8. Explorer-1
Figure 1.9. Vanguard-1 (TV-4)
Figure 1.10. TIROS-1
Figure 1.11. Echo-1
Figure 1.12. Courier-1B
Figure 1.13. Telstar-1
Figure 1.14. SYNCOM-2
Figure 1.15. Intelsat-1
Figure 1.16. Molniya series satellite
Figure 1.17. Molniya orbit
Figure 1.18. Intelsat-3
Figure 1.19. Intelsat-4
Figure 1.20. Intelsat-5
Figure 1.21. Ekran series
Figure 1.22. Anik-A
Figure 1.23. Eutelsat-II
Figure 1.24. INSAT-4A
Figure 1.25. ESSA satellites
Figure 1.26. GOES satellite
Figure 1.27. METEOSAT series
Figure 1.28. INSAT series

Figure 1.29. Landsat-7

Figure 1.30. SPOT-5

Figure 1.31. Resourcesat

Figure 1.32. Molniya series

Figure 1.33. Proton series

Figure 1.34. Energia series

Figure 1.35. Delta series

Figure 1.36. Titan series

Figure 1.37. Space shuttle

Figure 1.38. Buran series

Figure 1.39. Ariane-5ECA

Figure 1.40. Long march

Figure 1.41. GSLV

Figure 2.1. A satellite reaching the orbit

Figure 2.2. Artist's image of Europe's launcher family

Figure 2.3. Paths of GEO, MEO, and LEO

Figure 2.4. Diagram of geosynchronous and geostationary orbits

Figure 2.5. Drawing of geostationary orbit

Figure 2.6. Comparison of medium earth orbital and low earth orbital

Figure 2.7. Drawing of low earth orbitals

Figure 2.8. Highly elliptical satellite orbit (HEO)

Figure 2.9. The angle of inclination of a satellite orbit

Figure 2.10. Launch and rising to space (yellow line) develops the GTO (geostationary transfer orbit) (blue line) when the rocket discharges the satellite in space on a path to geostationary orbit (red line)

Figure 2.11. The ESA telescope Gaia orbits around an L-point. The point is accurately behind the earth, so at this point, Gaia would be in the earth's shadow and incapable to receive the sunlight required to power its solar panels. Every few years, Gaia usages its motors to regulate its position to sustain this orbit

Figure 2.12. The Thuraya coverage area

Figure 3.1. A global LEOS communication system

Figure 3.2. Examples of satellite constellations

Figure 3.3. Example of a satellite along with its immediate eight inter-satellite links

Figure 3.4. Different satellite interconnection patterns

Figure 3.5. Worst-case delay. It is a function of satellite altitude and crosslink patterns

Figure 3.6. Spaced-based vs. ground-based routing

Figure 4.1. Medium earth orbit satellite constellation displaying customer and gateway spot beams

Figure 4.2. E-model R rating performance vs. one-way delay via ITU recommendation G

Figure 4.3. TCP alone amount in Mbps; 1500-byte packets, 50 Kbyte, and 1 Mbyte TCP window sizes, and 500 msec and 120 msec link round trip

Figure 4.4. TCP with PEP enabled throughput in Mbps; 1500-byte packets, 50 Kbyte and 1 Mbyte TCP window sizes, and 500 msec and 120 msec link round trip latencies

Figure 4.5. Dual antenna of the ground terminal

Figure 5.1. The delimited angle of the earth as understood from a satellite in the Clarke belt is approximately 17°

Figure 6.1. Schematic of a cold gas propulsion system

Figure 6.2. Schematic of a bi-propellant liquid propulsion system

Figure 6.3. Schematic of a solid propulsion system

Figure 6.4. Schematic of a Resistojet propulsion

Figure 6.5. Schematic of a radio frequency (RF) ion propulsion system

Figure 6.6. Schematic of a Hall effect propulsion system

Figure 7.1. SMART-1

Figure 7.2. NEOSSat

Figure 7.3. Characteristics of the nanosatellite

Figure 7.4. ELaNa

Figure 7.5. The three-dimensional model of the picosatellite

Figure 7.6. Next-generation of sensors for the femto-satellites

Figure 7.7. KickSat mission

Figure 7.8. Flight model configuration of Ten-Koh satellite in orbit. The black outer structure is formed of CFRP. (a) On the left is the computer-aided design and on right is the flight model photograph; (b) the envelope permitted by the launch vehicle was 500 ×500 × 500 mm

Figure 7.9. (a) System architecture of Ten-Koh system; (b) Ten-Koh payload architecture comprising: ECU (experiment control unit), MM (material mission), CPD (charged particle detector), and DLP (double Langmuir probe)

Figure 7.10. Philosophy of attitude determination for a Ten-Koh satellite. The magnetometer is utilized for space environment purposes along with ADS

Figure 7.11. Demonstration of the double Langmuir probes measuring system: (a) current-voltage sweep locations around the revolving spacecraft; and (b to d) one-dimensional model demonstrating the electrodes outside and inside of the plasma sheath

Figure 7.12. Demonstration of the double Langmuir probes measuring system: (a) double Langmuir probes electrode composed of solid aluminum, covered with 10 µm of Au in a spherical tip, the ceramic material in a shaft for separation, and FR4 plate on base for isolation purposes from the spacecraft's structure; and (b) the block diagram of double Langmuir probes measuring circuit. The dashed line specifies the double Langmuir probes measuring circuit physically employed in the single PCB

Figure 8.1. Array fed adaptive membrane reflector; untethered or tethered

Figure 8.2. Abstract scheme of a Pico satellite array free-flying in space

Figure 8.3. Diagrammatic scheme of satellite power system

Figure 8.4. Design ideas from Nasa for lightweight solar concentrators

Figure 8.5. Satellite ground systems

LIST OF TABLES

Table 6.1. Summary of cold gas propulsion systems
Table 6.2. Summary of liquid propulsion systems
Table 6.3. Summary of solid propulsion systems
Table 6.4. Summary of Resistojet propulsion systems
Table 6.5. Summary of RF ion propulsion systems
Table 7.1. Grouping of satellites centered on wet mass

LIST OF ABBREVIATIONS

ABI	Advanced Baseline Imager
AceS	Asian Cellular System
ACTS	Advanced Communications Technology Satellite
ADN	Ammonium Dinitramide
ADS	Attitude Determination Subsystem
AFRL	Air Force Research Laboratory
ANDE	Atmospheric Neutral Density Experiment
Ar	Argon
ASIC	Application Specific Integrated Circuit
AT&T	American Telephone and Telegraph
B2B	Business-to-Business
Bi	Bismuth
CD	Converging-Diverging
CDMA	Code Division Multiple Access
CGP	Cold Gas Propulsion
CHIPS	CubeSat High Impulse Propulsion System
COMM	Communication Subsystem
COMSAT	Communication Satellite Corporation
COTS	Commercial off-the-Shelf
CPD	Charged Particle Detector
CTE	Coefficient of Thermal Expansion
CubeSats	Cube-Satellites
DAC	Digital-to-Analog Converter
DMSP	Defense Meteorological Satellite Program
DSSP	Digital Solid-State Propulsion
ECS	European Communications Satellite
EGC	Enhanced Group Calls

EHF	Extremely High Frequency
EHIS	Energetic Heavy Ion Sensor
EIRP	Equivalent Isotropic Radiated Power
EPS	Electrical Power Subsystem
ESA	European Space Agency
ESP	Electric Solid Propellant
ESSA	Environmental Science Service Administration
EUVS	Extreme Ultraviolet Sensor
FCC	Federal Communications Commission
FMMR	Free Molecule Micro Resistojet
FY	Feng Yun
GCR	Galactic Cosmic Rays
GEM	Green Mono-Propellant
GEO	Geostationary Orbit
GFRP	Glass Fiber Reinforced Polymer
GIOVE-A	Galileo In-Orbit Validation Element-A
GLM	Geostationary Lightning Mapper
GMS	Geostationary Meteorological Satellite
GOES	Geostationary Operational Environmental Satellite
GPS	Global Positioning System
GTO	Geostationary Transfer Orbit
HAN	Hydroxyl Ammonium Nitrate
HAPS	High Altitude Platform Systems
HEO	Highly Elliptical Orbit
HSD	High-Speed Data
I	Iodine
IETF	Internet Engineering Task Force
IL	Ionic Liquid
INMARSAT	International Maritime Satellite Organization
INSAT	Indian Satellite
INTELSAT	International Telecommunications Satellite Organization
IOV	In-Orbit Validation
IP	Internet Protocol
IPoS	IP Over Satellite
IPSec	IP Security

IRS	Indian Remote Sensing Satellite
ISR	Intelligence, Surveillance, and Reconnaissance
ISS	International Space Station
ITU	International Telecommunication Union
Kr	Krypton
LANs	Local Area Networks
LEO	Low Earth Orbit
LEOS	Low Earth Orbit Satellites
LMDS	Local Multipoint Distribution System
LOS	Line of Sight
LP	Liquid Propulsion
MEO	Medium Earth Orbit
MEOP	Maximum Expected Operating Pressure
MES	Mobile Earth Station
METEOSAT	Meteorological Satellite
MIDAS	Missile Defense Alarm System
MIPS	Magnetospheric Particle Sensors
MISSE	Materials International Space Station Experiment
MM	Material Mission
MMDS	Multichannel Multipoint Distribution System
MS	Magnetic Shielding
MSS	Multispectral Scanners
NLV	Nanosatellite Launch Vehicle
NOAA	National Oceanic and Atmospheric Administration
OBC	Onboard Computer
PCS	Personal Communications System
PEP	Performance Enhancing Proxies
PV	Photovoltaic
R	Radius
RF	Radio Frequency
RIT	Radio-Frequency Ion Thruster
RRS	Radio Relay Satellites
RSS	Remote Sensing Satellites
SAR	Synthetic Aperture Radar
SEE	Single Event Effects

SEISS	Space Environment *In-Situ* Suite
SEP	Solar Energetic Particle
SGPS	Solar and Galactic Proton Sensor
SIS	Solar Imaging Suite
SPMs	Solid Propellant Micro-Thrusters
SPOT	Satellite Pour L'observation De La Terre
SPS	Solar Power Satellites
SRP	Solid Rocket Propulsion
SSIs	Solid State Ionics
SSO	Polar Orbit and Sun-Synchronous Orbit
SSTL	Surrey Satellite Technology Ltd.
SUVI	Solar Ultraviolet Imager
TCP	Transmission Control Protocol
TDMA	Time Division Multiple Access
TIROS	Television and Infrared Observation Satellite
TM	Thematic Mappers
UAVs	Unattended Aerial Vehicles
UN	United Nations
VHRR	Very High-Resolution Radiometers
VPN	Virtual Private Network
Xe	Xenon
XRS	Solar X-Ray Sensor

PREFACE

Satellite technology is a popular field in today's world. It seems quite familiar to many of us regardless of our professional and academic backgrounds. This technology is no longer a prerogative of particular countries and is not confined to the research labs of only big research organizations and academic institutes. Presently, satellite technology is taught as one of the primary courses at undergraduate and postgraduate levels. Most of the literature on satellite technology encompasses the applications of satellites in communication systems only. The information about other aspects of satellite technology is usually overlooked, which include its applications in weather forecasting, remote sensing, scientific research, navigational applications, and military uses.

This book, *Introduction to Satellite Technology and Its Applications*, is a comprehensive book on the topic of contemporary satellite technologies and their applications. Offering a widespread overview of applications, from military uses and remote sensing to scientific and navigational applications, the book also presents inclusive information on different satellite types of space launch vehicles.

There are eight chapters in the book. Each chapter offers a thorough introduction of a particular topic related to satellite technology. Chapter 1 introduces the readers to the fundamentals of satellites and their types. Chapter 2 focuses on the discussion of different types of satellites and their basic classifications. Different types of orbits are also discussed in the chapter.

Communication systems based on low earth orbit satellites (LEOS) are an exciting and fascinating endeavor in reorganizing the services that the global communication network provides. Chapter 3 offers key information regarding LEOS systems and their applications. Chapter 4 describes the design of an effective and economlow-latencytency MEO constellation by the usage of high-power dirigible Ka-band spot beams. Chapter 5 illustrates the fundamentals of geosynchronous satellites. The past few years have observed an incredible rise in curiosity in the use of cube-satellites (CubeSats) among the space community comprising space agencies, the industry as well as academia. Chapter 6 offers a thorough overview of cube-satellite propulsion technologies.

A miniaturized satellite, small satellite, or simply small-sat is a satellite having a low size and mass, generally less than 500 kg. Chapter 7 provides an overview of different small satellites and their applications. Finally, Chapter 8 contains information about future trends in satellite technology and satellite communication systems.

The book can serve as an ideal guide for researchers and professionals in the field of satellite technology and space sciences. Moreover, engineering students from multidisciplinary fields can also benefit from the book.

Chapter

1

INTRODUCTION TO SATELLITES AND THEIR APPLICATIONS

CONTENT

1.1. Introduction ... 2

1.2. What is the Satellite? ... 3

1.3. The Era of Sounding Rockets and Hot Air Balloons ... 6

1.4. Launch of the Early Artificial Satellites ... 7

1.5. Satellites For Meteorology, Scientific Exploration, and Communications: Early Developments ... 10

1.6. Non-Geosynchronous Communication Satellites: Relay and Telstar Programs ... 11

1.7. Development of the Geosynchronous Communication Satellites ... 13

1.8. Global Communication Satellite Systems ... 15

1.9. Local Communication Satellite Systems ... 18

1.10. Satellites for Various Other Applications Also Made Quick Progress ... 21

1.11. Evolution of the Launch Vehicles ... 26

1.12. Future Trends ... 32

References ... 35

1.1. INTRODUCTION

Nowadays, the term 'Satellite' is a very familiar name. It sounds so acquainted to everyone regardless of professional and educational background. It is the subject of discussion and interest not only to communication and electronics engineers, technocrats, and scientists; it fascinates electronics enthusiasts, hobbyists, and almost everyone (Elbert, 2008; Huang et al., 2016).

Different stages of advancement of satellite launch vehicles and satellites will be discussed briefly in this chapter, starting with the era of sounding rockets and hot air balloons of 1940–1950 to the modern-day status (Gan and Zhang, 2006; Li et al., 2016).

The prospect of applications of satellite has prolonged far beyond offering intercontinental satellite television and communication services. Space exploration and atmospheric monitoring are the other main frontiers where satellite utilization has been used a great deal. Then comes the host of defense-associated applications, which comprise secure communications, spying, navigation, and so on (Lutz et al., 1991; Bacsardi, 2013).

The application areas are increasing, and so is the significance of applications in those areas. For example, in the area of communication-associated applications, it is not just the long-distance facsimile and telephony and video services that are significant; satellites are generally playing a growing role in the newer communication services like mobile communication, data communication, and then so on (Arnon and Kopeika, 1997; Wang *et al.*, 2019). Video conferencing, where diverse people at dissimilar locations, no problem how far away these locations are, can conduct meetings in real-time in order to exchange thoughts or make significant decisions, is the reality nowadays in big establishments. The Internet and the innovative services it provides are well-known to all. Satellites are the foundation of all these happenings (Sharma *et al.*, 2013; Rahmat-Samii and Densmore, 2014).

The satellite is frequently known as an 'orbiting radio star' for reasons that can be appreciated easily. These orbiting radio stars help aircraft and ships to navigate safely in all conditions of weather. It is fascinating to learn that some classes of medium-long-range cruise and ballistic missiles need the aid of the satellite to hit their planned targets accurately. The satellite-based GPS (global positioning system) is utilized as assistance to navigate securely and safely in unknown territories (Kuang et al., 2017; Petrov *et al.,* 2017). Remote sensing and earth observation satellites provide information regarding the weather, volcanic eruptions, ocean conditions, pollution, earthquakes, and the health of forests and crops. Another category of satellites keeps an eye on military activity all over the globe and aids to some extent in policing or enforcing arms control agreements (Misra *et al.,* 2013; Yan *et al.,* 2019).

Even though mankind is still to tour beyond the moon, satellites have traversed the solar system in order to explore all planets. The satellites with astrophysical applications have huge telescopes and have sent information that has triggered many novel discoveries, throwing novel light on the universe. Due to these reasons, almost all the developed nations comprising the United Kingdom, the United States, France, Germany, Japan, and Russia, and developing countries such as India have the heavily funded and fully-fledged space program, controlled by organizations with substantial technical and scientific manpower and infrastructure (Sadek and Aissa, 2012; Riva *et al*., 2014).

1.2. WHAT IS THE SATELLITE?

Generally, a satellite is an artificial or natural body moving around the celestial body like a star or a planet. In the current chapter, reference is made just to artificial satellites moving around the planet Earth. The satellites are placed into the anticipated orbit and have different payloads dependent upon the desired application (Fernández, 2011; Radhakrishnan *et al*., 2016).

The concept of the geostationary satellite was instigated in father paper published by a science fiction writer, Arthur C. Clarke, in the magazine *Wireless World* in 1945. In that paper, he stressed the significance of the orbit whose radius (R) from Earth's center was such that the period of orbit was equal to the time taken by the planet Earth to finish one rotation around the axis (Hnatushenko *et al.,* 2016). Though the notion of the satellite initiated from an aspiration to place an object in space that would seem to be still with respect to the surface of the Earth, therefore making possible the presenter of communication services, several other diversities of the satellites also exist where they require not to be still with respect to the observer on Earth in order to perform the anticipated function (Block et al., 1998; Livingston and Belle, 2005).

The satellite in orbit executes its selected role over its lifetime. The communication satellite (Figure 1.1) is generally a type of repeater station that obtains signals from the Earth, processes the signals, and retransmits these signals back to the Earth. The Earth observation satellite (Figure 1.2) has a camera as the payload in order to take pictures of the interesting regions during the periodic motion. The weather forecasting satellite (Figure 1.3) takes pictures of clouds and also observes other atmospheric parameters, therefore helping the weather forecast department in making accurate and timely forecasts (Comparetto and Ramirez, 1997; Underwood *et al*., 2015).

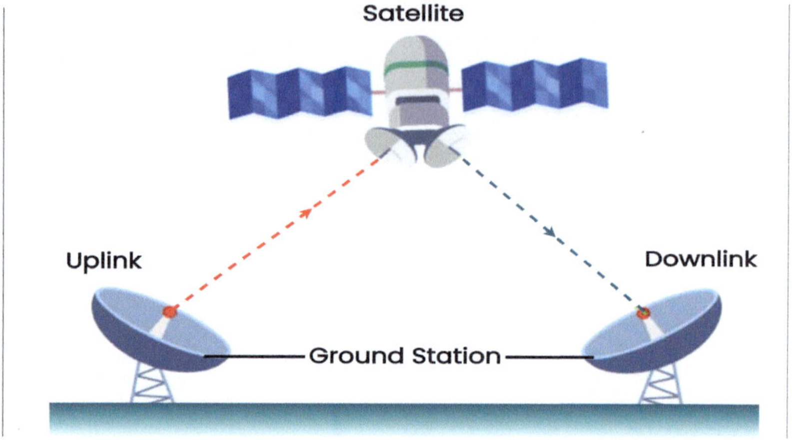

Figure 1.1. *A communication satellite.*

Source: https://byjus.com/physics/satellite-communication/.

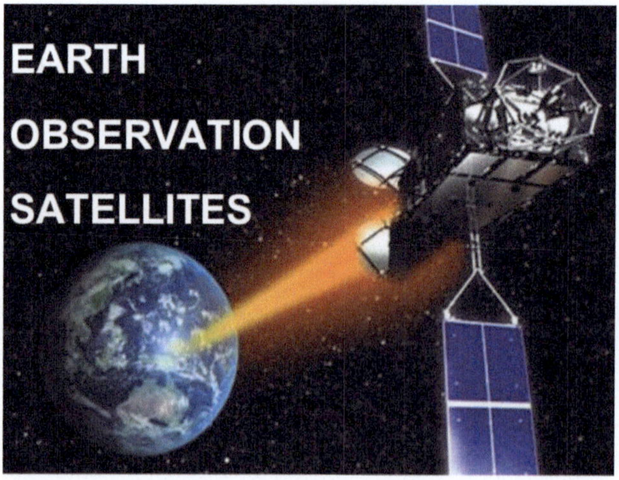

Figure 1.2. *An earth observation satellite.*

Source: https://www.slideshare.net/dbasuti/earth-observation-satellites-13 031 6930.

Figure 1.3. *Weather forecasting satellite.*

Source: https://en.wikipedia.org/wiki/Weather_satellite.

The satellite could efficiently play the role of a spy in a situation of some military-purpose-built satellites (Figure 1.4) or can also do the job of an explorer when appropriately prepared and launched for astrophysical applications (Figure 1.5).

Figure 1.4. *United States military satellite.*

Source: https://www.intelligent-aerospace.com/satcom/article/14073193/us-military-satellite-warfare.

Figure 1.5. *Scientific satellite.*

Source: https://en.wikipedia.org/wiki/Satellite.

It all started with the article issue of *Wireless World* published in October 1945 by Arthur C. Clarke, which theoretically suggested the possibility of putting a communication satellite in the geostationary orbit (GEO). In the article, he discoursed how the GEO satellite would appear stationary to a spectator on Earth inside the coverage of the satellite, therefore providing a continuous communication service all over the world. This marked the commencement of the satellite period. Technologists and scientists started to look earnestly at such a likelihood and the breakthrough it was possible to bring along with such a likelihood (Hall, 1988; Esper *et al.*, 2000).

1.3. THE ERA OF SOUNDING ROCKETS AND HOT AIR BALLOONS

The implementation of the mission started with the introduction of sounding rockets and hot air balloons utilized for the aim of aerial observation of Earth normally from the upper reaches of the atmosphere of Earth. These sounding rockets carried a diversity of instruments to execute their relevant mission objectives (Kessler and Cour - Palais, 1978; Kankaku *et al.*, 2014).

A-4 (V-2) rockets utilized broadly during World War II for transporting explosive warheads engrossed the consideration of users of sounding rockets for the aim of scientific exploration of the upper atmosphere with the help of the high-altitude rocket. The first of the A-4 rocket to carry the scientific instruments to the upper atmosphere was lifted off in May 1946 (Figure 1.6). The rocket took an instrument with it to record the cosmic ray flux from an altitude of almost 112 km. This launch was trailed by various more during that particular year (Renzetti, 2013; Belward and Skøien, 2015).

The final V-2A rocket (Soviet version of altered A-4 rocket), made its entrance in 1949. It carried a payload of nearly 860 kilograms and attained a height of almost 212 km (Yajima *et al.*, 2009; Unwin *et al.*, 2016).

Figure 1.6. *1st A-4 rocket to be taken off.*

Source: Image courtesy: NASA; https://www.nasa.gov/multimedia/imagegallery/image_feature_644.html.

1.4. LAUNCH OF THE EARLY ARTIFICIAL SATELLITES

The Russia and United States were the first countries to develop plans for artificial satellites in the year 1955. Both countries proclaimed their proposals to build and lift off artificial satellites. Within only 2 years, Russians achieved the feat and the US followed rapidly thereafter (Sweeting, 2018).

Sputnik-1 (Figure 1.7) was the 1st artificial satellite that gave life the age of space exploration. Launched by Soviet R7 ICBM from Baikonur Cosmodrome on 4 October 1957, it orbited the Earth every 96 mins in the elliptical orbit of 227 kilometers× 941 kilometers inclined at 65.1° and was developed to give information on the temperature and density of the upper atmosphere. After the 92 fruitful days in orbit, Sputnik-1 burned as the satellite fell from the orbit into an atmosphere in January 1958 (Sellers *et al.*, 1992).

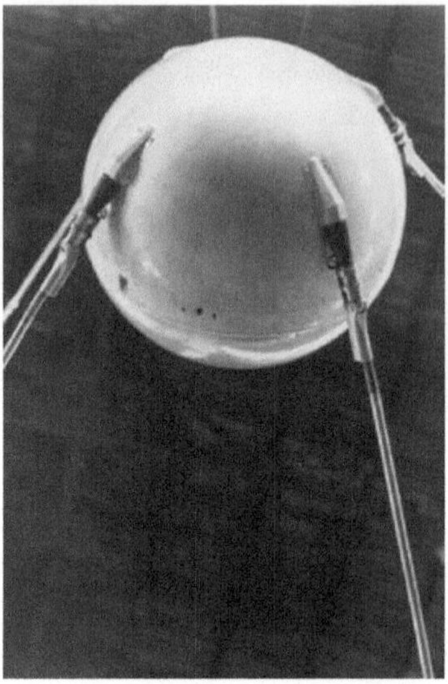

Figure 1.7. *Sputnik-1.*

Source: Image courtesy: NASA; https://www.nasa.gov/mission_pages/explorer/sputnik-20071002.html.

Sputnik-2 and Sputnik-3 trailed Sputnik-1. On 3 November 1957 Sputnik-2 was launched in the elliptical orbit of 212×1660 kilometers inclined at 65.33°. Sputnik-2 carried a female dog called Laika, in flight. This female dog was the 1st living being to orbit Earth. This mission offered data on the biological effects of orbital flight. On 15 May 1958, a geophysical satellite named Sputnik-3 was launched that provided data on Earth's ionosphere, cosmic rays, meteoroids, and magnetic field. The orbital parameters of this satellite were 1864 km (apogee), 217 km (perigee), and 65.18° (orbital inclination).

The launches of satellites Sputnik-1 and Sputnik-2 had embarrassed and surprised the Americans as till that date they had no fruitful satellite launch. The 1st satellite to be launched successfully by the US was Explorer-1 (Figure 1.8). On 31 January 1958, Explorer-1 was launched from Cape Canaveral by Jupiter-C rocket. The orbital parameters of this satellite were 2534 km (apogee), 360 km (perigee), and 33.24° (orbital inclination). The design of this satellite was pencil-shaped, which permitted it to spin just like a bullet as it orbited around the Earth. The spinning motion gave stability to Explorer-1 while in orbit. Parenthetically, spin stabilization is amongst the conventional methods of satellite stabilization (Rahmat-Samii *et al.*, 2017).

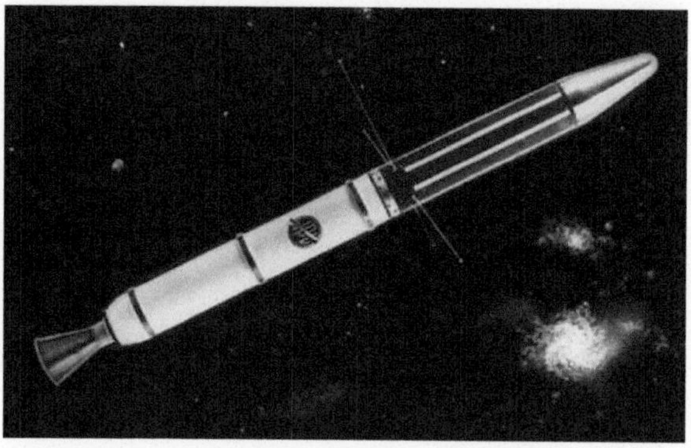

Figure 1.8. *Explorer-1.*

Source: Image courtesy: NASA/JPL-Caltech; https://www.jpl.nasa.gov/news/explorer-1-the-beginning-of-american-space-science.

After the fruitful liftoff of Explorer-1, there tailed in quick sequence the liftoffs of Vanguard-1, Explorer-2, and Vanguard-1 (TV-4) during February-March 1958 (Figure 1.9). The launches of Explorer-2 and Vanguard-1 were unsuccessful. The launch of Vanguard-1 (TV-4) was successful. Vanguard-1 (TV-4) was the 1st satellite to engage solar cells to charge the satellite batteries. The satellite's orbital parameters were 2465 km (apogee), 404 km (perigee), and 34.25° (orbital inclination). This mission performed geodetic studies and discovered that the Earth was pear-shaped.

Figure 1.9. *Vanguard-1 (TV-4).*

Source: Image courtesy: NASA; https://nssdc.gsfc.nasa.gov/nmc/spacecraft/display.action?id=1958-002B.

1.5. SATELLITES FOR METEOROLOGY, SCIENTIFIC EXPLORATION, AND COMMUNICATIONS: EARLY DEVELOPMENTS

American experiences with liftoffs of the Explorer and Vanguard series of satellites and Soviet experiences with launches of Sputnik series had taken satellite launch and satellite technology to adequate maturity. The period between 1960 and 1965 saw the liftoffs of experimental satellites launched for the applications mentioned above. The year 1960 was quite busy for the purpose. This year witnessed the successful liftoffs of the 1st weather satellite TIROS-1 (television and infrared observation satellite) (Figure 1.10) on 1st April 1960, MIDAS-2 the 1st experimental infrared surveillance satellite in May 1960, Echo-1 (Figure 1.11), the 1st experimental passive communication satellite on August 1960 and Courier-1B (Figure 1.12) the active repeater communication satellite in October 1960. Additionally, that year also witnessed fruitful liftoffs of satellites Sputnik-5 and Sputnik-6 in August and December correspondingly (Woods et al., 2010; Ma *et al.*, 2019).

Figure 1.10. *TIROS-1.*

Source: Image courtesy: NASA; https://www.nesdis.noaa.gov/content/celebrating-world%E2%80%99s-first-meteorological-satellite-tiros-1.

The Echo sequence of satellites, which usually were aluminized Mylar balloons behaving as passive reflectors, developed how two remotely situated stations on Earth might communicate with one another through the space-borne passive reflector. The missile defense alarm system (MIDAS) series of timely warning satellites developed beyond any uncertainty the significance of surveillance from the space-borne platforms in order to locate and recognize the program of strategic weapon development of an adversary. The satellites Sputnik-5 and Sputnik-6 further examined the biological effects of orbital flights. Both spacecraft had carried two dog passengers (Kramer and Cracknell, 2008).

Figure 1.11. *Echo-1.*

Source: Image courtesy: NASA; https://history.nasa.gov/SP-4217/ch4.htm.

Figure 1.12. *Courier-1B.*

Source: Image courtesy: US Army; https://en.wikipedia.org/wiki/Courier_1B.

1.6. NON-GEOSYNCHRONOUS COMMUNICATION SATELLITES: RELAY AND TELSTAR PROGRAMS

Having developed the concept of active and passive repeater stations to the relay communication signals, the next significant phase in the history of the satellite was the utilization of non-geostationary satellites for the intercontinental communication services. The procedure was initiated by AT&T (American Telephone and Telegraph) seeking approval from FCC (Federal Communications Commission) to launch the experimental communications satellites. The Relay sequence of satellites that trailed the Telstar also has an association to the same category (Ward, 1993; Karafantis, 2016).

In the series of Telstar, Telstar-1 (Figure 1.13), the 1st communication satellite and the 1st commercially sponsored satellite, was lifted off in July 1962, trailed a year later by the Telstar-2 in May 1963. The satellite Telstar-2 had a higher orbit in order to decrease exposure to the harmful effect of the radiation belt. The satellite Telstar-1 with the orbital parameters of 5632 km (apogee), 952 km (perigee), and an orbital inclination of 44.79° started the innovation in global television communication from the non-geosynchronous orbit. It connected Europe and the United States (Bodroghkozy, 2016).

Figure 1.13. *Telstar-1.*

Source: Image courtesy: NASA; https://www.nasa.gov/topics/technology/features/telstar.html.

Telstar-1 was tailed by Relay-1 (a prototype of the operational communication satellite by NASA) launched in December 1962. The next satellite in the series, Relay-2 was launched in January 1964. The satellite's orbital parameters were 7439 km (apogee), 1322 km (perigee), and 47.49°

(inclination). The objectives of the mission were to trial the transmissions of telephone, television, digital data, and facsimile (Bhasin *et al.*, 1998). It is worth mentioning here that the Relay and Telstar sequence of satellites were the experimental vehicles developed to discover the boundaries of the performance of the satellite and were just the introduction to bigger events to tail.

1.7. DEVELOPMENT OF THE GEOSYNCHRONOUS COMMUNICATION SATELLITES

The subsequent main milestone in satellite technology history was the idea of Arthur C. Clarke becoming a reality. The small-scale spin-stabilized satellite was shown first in 1961 at Paris Air Show. In February 1963, SYNCOM-1 was launched, but shortly after, the mission was unsuccessful. SYNCOM-2 (Figure 1.14), launched in July 1963, became the 1st functional geosynchronous communication satellite. This satellite was then followed by SYNCOM-3 (Wrenn, 1995; Muri and McNair, 2012).

Figure 1.14. *SYNCOM-2.*

Source: Image courtesy: NASA; https://science.howstuffworks.com/satellite7.htm.

This satellite was put over the equator close to the international date line in August 1964. SYNCOM-3 was utilized to broadcast the opening ceremonies of the Tokyo Olympics. The world started to witness the words 'live via satellite' for the first time on the television screens (Fugono et al., 1980; Kumar and Kumar, 2001).

One more important development during that time was the creation of the International Telecommunications Satellite Organization (INTELSAT) in August 1964 with Communication Satellite Corporation (COMSAT) as its operational support. INTELSAT accomplished the main milestone with the liftoff of the Intelsat-1 satellite, known as Early Bird (Figure 1.15), from Cape Canaveral in April 1965.

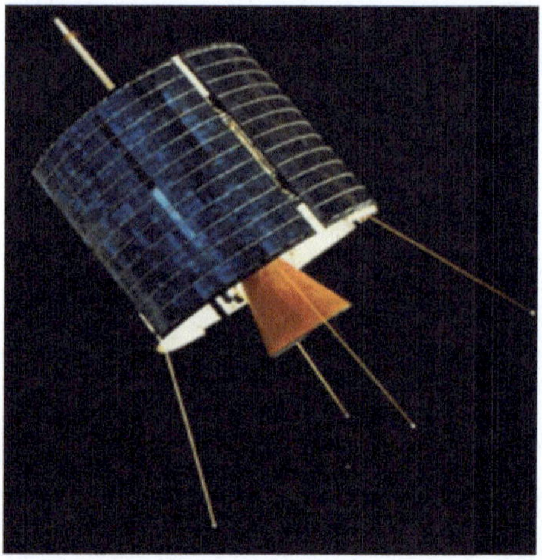

Figure 1.15. *Intelsat-1.*

Source: Replicated by approval of © Intelsat; https://space.skyrocket.de/doc_sdat/intelsat-1.htm.

Intelsat-1 was the 1st geostationary communication satellite in a commercial facility. It went into the systematic facility in June 1965 and offered 240 telephone circuits for the connectivity between North America and Europe. Although developed for the anticipated life span of just 18 months, it stayed in the facility for more than 3 years (Janson, 2020).

Whereas the Americans developed their competence in launching the communications satellites through liftoffs of the SYNCOM sequence of satellites and Intelsat-1 satellite during the era of 1960–1965, the Soviets did this with the help of their Molniya sequence of satellites starting April 1965. The Molniya sequence of satellites (Figure 1.16) were exclusive in providing continuous communications services 24 hours a day without being in a conventional GEO. The satellites followed elliptical and highly inclined orbits, called the Molniya orbit (Figure 1.17), with perigee and apogee distances of nearly 500 km and 40,000 km and an inclination of 65°.

Figure 1.16. *Molniya series satellite.*

Source: https://space.skyrocket.de/doc_sdat/molniya-1s.htm.

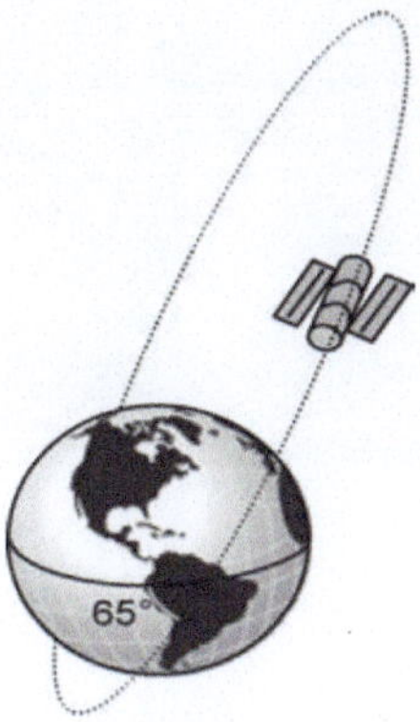

Figure 1.17. *Molniya orbit.*

Source: https://en.wikipedia.org/wiki/Molniya_orbit.

Satellites in such orbit with an orbital period of 12 hours stayed over the countries of the previous Soviet Union in the northern hemisphere for nearly more than 8 hours. The series of Molniya-1 was tailed later by a Molniya-2 in 1971 and then Molniya-3 series in 1974.

1.8. GLOBAL COMMUNICATION SATELLITE SYSTEMS

The Intelsat-1 (Early Bird) satellite was tailed by an Intelsat-2 sequence of satellites. Four Intelsat-2 satellites were sent into orbit within 1 year

(1966–1967). The next main landmark *vis-a-vis* communications satellites were accomplished with the Intelsat-3 sequence of satellites (Figure 1.18) becoming completely functional. The other novel concept tried fruitfully with the satellites was the utilization of the de-spun structure of the antenna, which permitted the utilization of a highly directional antenna on the spin-stabilized satellite. Intelsat-1 and the Intelsat-2 series of satellite utilized omnidirectional antennas (Zheng *et al.,* 2012; Kourogiorgas et al., 2017).

Figure 1.18. *Intelsat-3.*

Source: Replicated by approval of © Intelsat; https://space.skyrocket.de/doc_sdat/intelsat-3.htm.

The capabilities of communication satellites continued to upsurge with almost every novel venture. With the satellites of Intelsat-4 (Figure 1.19), the 1st of which was sent into orbit in 1971, the capacity of the satellite got a big boost. The features of frequency re-utilization were taken to an additional dimension in the series of Intelsat-5 with the utilization of polarization discrimination. Whereas frequency re-utilization, i.e., utilization of the same frequency band, was feasible when two footprints were spatially away from each other, dual-polarization permitted the re-utilization of the same frequency band inside the same footprint. The satellites of Intelsat-5 (Figure 1.20), the 1st of which was sent into orbit in 1980, utilized both Ku band and C band transponders and generally was 3-axis stabilized. The series of Intelsat-9 and Intelsat-10 were sent into orbit in the first era of the new millennium (Grami, 2006; Golkar and iCruz, 2015).

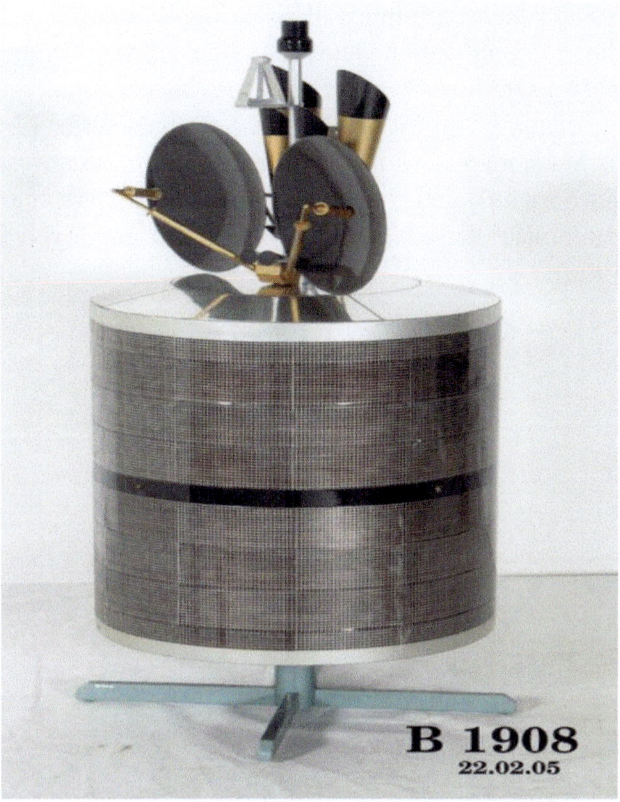

Figure 1.19. *Intelsat-4.*

Source: Replicated by approval of © Intelsat; https://collection.maas.museum/object/209263.

Figure 1.20. *Intelsat-5.*

Source: Replicated by approval of © Intelsat; https://space.skyrocket.de/doc_sdat/intelsat-5a.htm.

The Russians have continued their struggle towards designing and launching the communication satellites after accomplishment with the series of Molniya satellites. The Ekran series (internationally designated as Statsionar-T), displayed in Figure 1.21, the Raduga series (internationally designated as Statsionar-1), Molniya series, Gorizont series (internationally designated as horizon), and the Ekspress-AM series are the most recent Russian communication satellites. All three are employed in the GEO (Purchase, 1996).

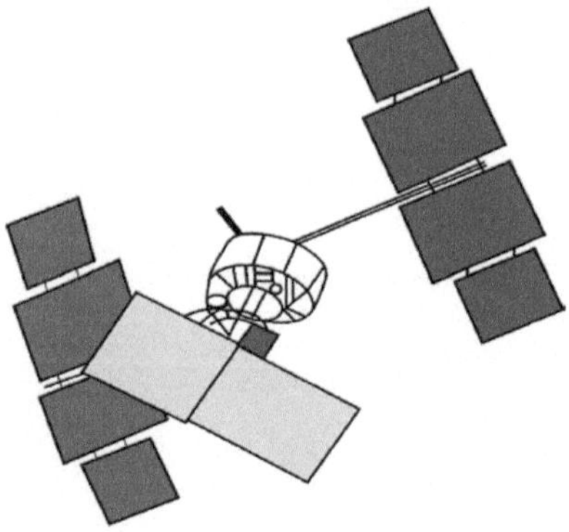

Figure 1.21. *Ekran series.*

Source: https://rammb.cira.colostate.edu/dev/hillger/Ekran.htm.

1.9. LOCAL COMMUNICATION SATELLITE SYSTEMS

Starting in 1965, the series of Molniya satellites established the helpfulness of the local communications satellite systems when it offered communications connectivity to the large number of republics distributed over the massive landmass of the previous Soviet Union.

Such kind of system was mainly attractive for countries having enormous territory. Amongst the non-Soviet countries, Canada was the 1st one to have the dedicated national satellite system with the liftoff of the Anik-A series (Figure 1.22), starting in 1972. The abilities of the satellites were consequently augmented with the consecutive Anik series, named Anik-B (starting 1978), Anik-C (starting 1982), Anik-D (starting 1982), Anik-E (starting 1991), Anik-F (starting 2000), and Anik-G (starting 2013). Anik-G1, 1st satellite of this series, is the multi-mission satellite developed to offer DTH (direct-to-home television services in Canada and broadband video, data, and voice services in South America. On April 16, 2013, the satellite was put into orbit by Breeze-M rocket (Kudsia *et al*., 1992).

The US started its campaign for making national communication satellite systems with the liftoff of Satcom satellite in the year 1975, Comstar satellite in the year 1976, and Westar satellite in the year 1974. These satellites were trailed by various more ventures. Europe started with the ECS series (European communications satellite) and tailed it with the series of Eutelsat-II satellites (Figure 1.23) and the series of Eutelsat-W satellites. In the year 2012, EUTELSAT retitled all of its satellite series under the new name of Eutelsat (Greenberg, 1979; Evans, 2001).

Amongst the developing nations, Indonesia was the 1st to identify the potential of the national communication satellite systems and had the Palapa satellites launched in the year 1977 to connect her dispersed island nation. The series of Palapa satellites have seen four generations so far named Palapa-A (starting 1977), Palapa-B (starting 1984), Palapa-C (starting 1991), and Palapa-D (starting 2009). On August 31, 2009 Palapa-D was put into orbit with the help of the Chinese Long March 3B rocket (Kobayashi, 1995).

Figure 1.22. *Anik-A.*

Source: Image courtesy: Telesat Canada; https://www.ieee.ca/millennium/anik/anik_model.html.

China, India, Saudi Arabia, Mexico, Japan, and Brazil tailed suit with their particular local communication satellite systems. India started with the series of INSAT-1 satellites in the year 1981 and has entered the 4th generation with the series of INSAT-4. INSAT-4CR (Figure 1.24) was put into orbit in September 2007.

Figure 1.23. *Eutelsat-II.*

Source: Replicated by approval of © Eutelsat; https://www.eutelsat.com/en/group/our-history.html.

The newest in the series of INSAT-4 is the INSAT-4G satellite, put into orbit on 21 May 2011 by the Ariane-5 rocket. Nonetheless, the newest in the series of INSAT-3 is the INSAT-3D put into orbit on 26 July 2013 with the help of the Ariane-5 rocket. Arabsat, which connects the countries of the Arab League, has entered the 5th generation with the series of Arabsat-5 satellites.

Figure 1.24. *INSAT-4A.*

Source: Image courtesy: ISRO; https://alchetron.com/INSAT-4A.

1.10. SATELLITES FOR VARIOUS OTHER APPLICATIONS ALSO MADE QUICK PROGRESS

The aim to utilize satellites for applications instead of communications was very apparent, even in the initial stages of satellite development.

Numerous satellites were launched primarily by the United States and the former Soviet Union for navigation, surveillance, Earth observation, and meteorological studies during the 1960s (Price *et al.,* 2020).

Making the modest commencement with the series of TIROS, meteorological satellites (METEOSAT) have come quite a long way in terms of satellites put into orbit for the aim and also developments in the sensor's technology utilized on these satellites.

Main non-geostationary weather satellites that have progressed over the years comprise the TIROS series and the series of Nimbus starting around 1960, Environmental Science Service Administration (ESSA) series (Figure 1.25) starting in 1966, National Oceanic and Atmospheric Administration (NOAA) series starting in 1970, defense meteorological satellite program (DMSP) series started in 1965 (from the US), the Meteor series starting in the year 1969 from Russia and the Feng Yun series (FY-1 and FY-3) starting in 1988 from China.

Figure 1.25. *ESSA satellites.*

Source: Image courtesy: NASA; https://en.wikipedia.org/wiki/ESSA-8.

The main METEOSAT included in the geostationary group comprise the GOES satellite (Image courtesy: NASA and NOAA) geostationary meteorological satellite (GMS) series from Japan ever since 1977, geostationary operational environmental satellite (GOES) series from the US (Figure 1.26) ever since 1975, the METEOSAT series from Europe ever since 1977 (Figure 1.27), the Indian satellite (INSAT) series from India ever since 1982 (Figure 1.28) and the series of Feng Yun (FY-2) from China ever since 1997 (Kogan, 1995; Yang *et al*., 2018).

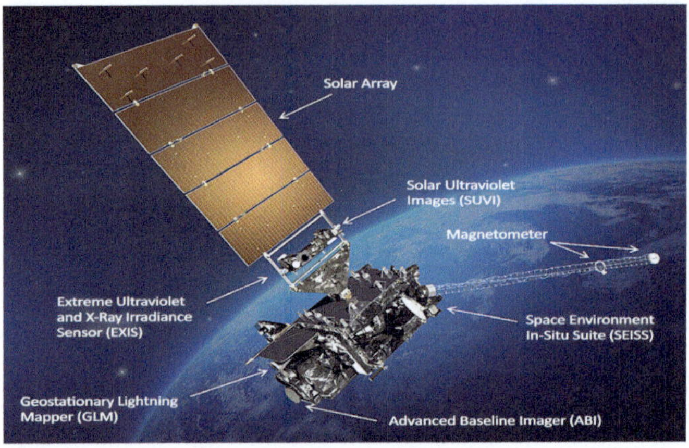

Figure 1.26. *GOES satellite.*

Source: Image courtesy: NASA and NOAA; https://directory.eoportal.org/web/eoportal/satellite-missions/content/-/article/goes-r.

Sensors utilized on these types of satellites have seen various technological advances, in types and the numbers of sensors utilized along with their levels of performance.

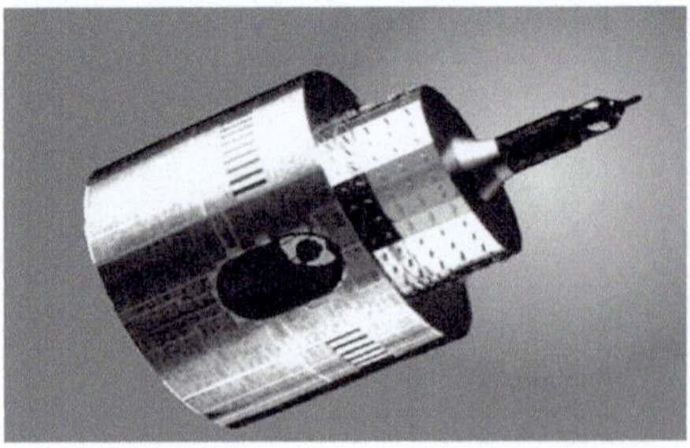

Figure 1.27. *METEOSAT series.*

Source: Replicated by approval of © EUMETSAT; https://en.wikipedia.org/wiki/Meteosat.

The satellites give very high-resolution pictures of Earth and cloud cover in infrared and visible parts of the spectrum and therefore help produce data on tropical storms, cloud formation, snow cover, hurricanes, temperature profiles, probability of forest fires, and so on (Iverson *et al.*, 1989).

RSS (remote sensing satellites) have come quite a long way ever since the initial 1970s with the liftoff of the 1st of the Landsat satellites series that gave comprehensive attention to several aspects of perceiving Earth from the space-based platform. The initial notions of having satellite systems for this aim came from the early black and white television pictures of Earth underneath the cloud cover as provided by TIROS weather satellite in 1960, tailed by splendid observations discovered by Astronaut Gordon Cooper in 1963 during his flight in the Mercury capsule when he discoursed to have seen buildings, roads, and smoke coming out of the chimneys from the altitude of almost 160 km. His discoveries were consequently confirmed during consecutive exploratory space missions (Koike, 1991).

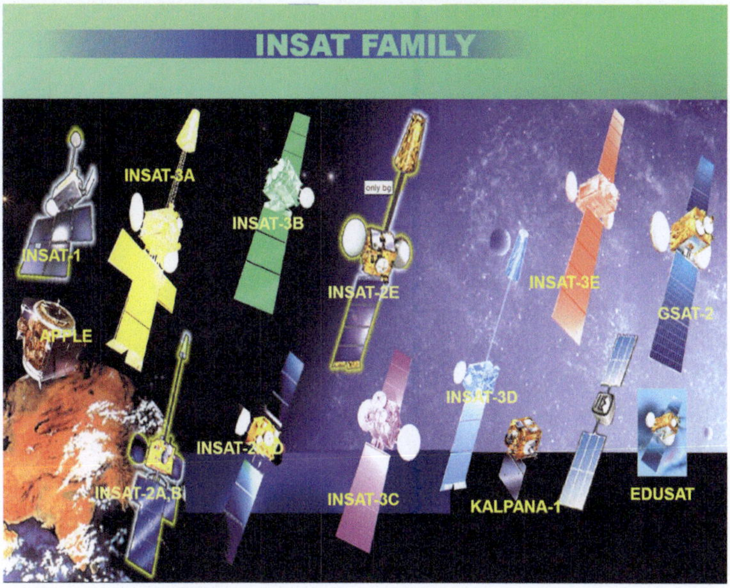

Figure 1.28. *INSAT series.*

Source: https://slideplayer.com/slide/12713121/.

With the passage of years, with substantial advances in several technologies, the applications of or RSS or spectrum of the Earth observation satellites has expanded very quickly from simple terrain mapping known as cartography to anticipating agricultural crop production, forestry, pollution monitoring, oceanography, and ice reconnaissance.

The Landsat program, starting with Landsat-I in the year 1972, has advanced to Landsat-8 over Landsat-2, -3, -4, -5, -6, and -7 (Figure 1.29). Landsat-8, the current in the series of Landsat, was put into orbit on 11 February 2013.

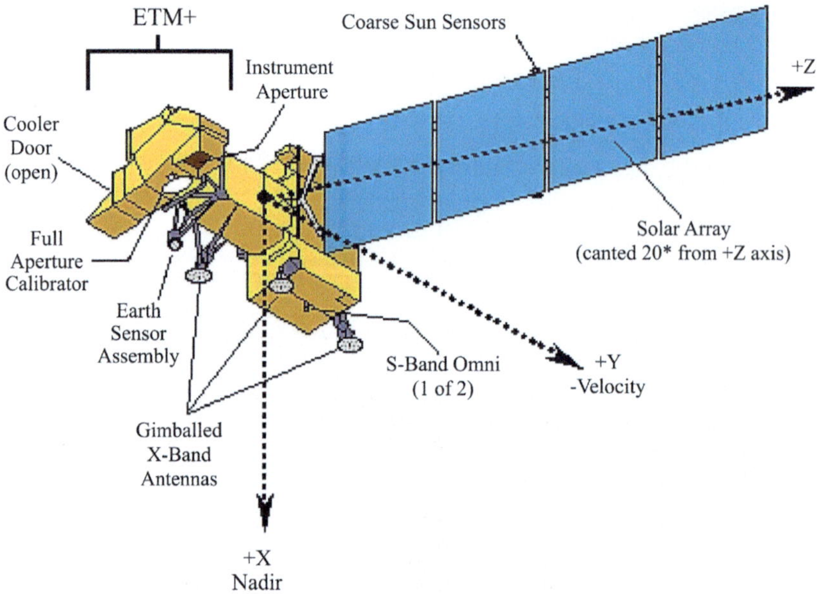

Figure 1.29. *Landsat-7.*

Source: Image courtesy: NASA; https://directory.eoportal.org/web/eoportal/satellite-missions/l/landsat-7.

The series of SPOT has come quite a long way, starting with SPOT-1 in the year 1986 to SPOT-6 put into orbit in 2012 over SPOT-2, -3, -4, and -5 (Figure 1.30).

The series of IRS launches started in the year 1988 with the liftoff of IRS-1A and the modern satellites put into orbit in this series are IRS-P6 known as Resourcesat 1 (Figure 1.31) launched in the year 2003 and the IRS-P5 known as Cartosat launched in the year 2005. Cartosat2, Cartosat-2A, Cartosat-2B, and the Resouresat 2 put into orbit in 2007, 2008, 2010, and 2011 correspondingly are other RSS of India. The sensors onboard the latest

Earth observation satellites comprise high-resolution television cameras, MSS (multispectral scanners), VHRR (very high-resolution radiometers), TM (thematic mappers), and SAR (synthetic aperture radar). RISAT-1, put into orbit on April 26, 2012, is a current satellite whose all of the weather radar pictures will facilitate disaster and agriculture management (Heath et al., 1972).

Figure 1.30. *SPOT-5.*

Source: Replicated by approval of © CNES/ill.D.DUCROS, 2002; https://www.satimagingcorp.com/satellite-sensors/other-satellite-sensors/spot-5/.

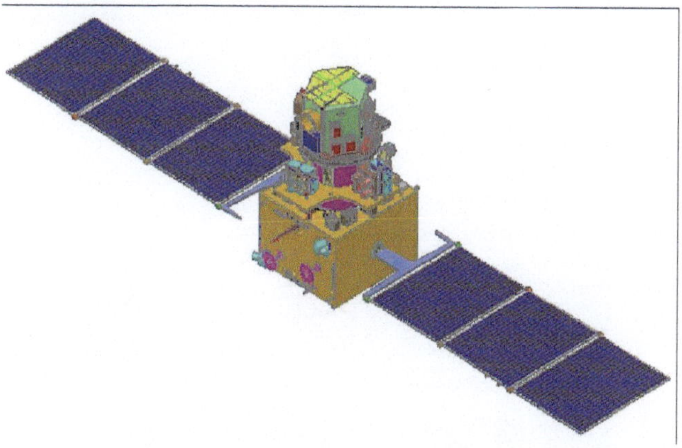

Figure 1.31. *Resourcesat.*

Source: Image courtesy: ISRO; https://directory.eoportal.org/web/eoportal/satellite-missions/r/resourcesat-2.

1.11. EVOLUTION OF THE LAUNCH VEHICLES

The launch vehicles have seen several stages of evolution to meet the launch demands of diverse classes of satellites. The necessity to make launch vehicles by the countries like Russia and the US was in the former stages targeted to attain technological superiority in space technology (Nagata et al., 1976; Takahashi, 2013). This led these countries to utilize the missile technology made during the World War II era to make launch vehicles for satellites. The next stage was to revolutionize and enhance the technology to such an extent that the launch vehicles became economically feasible. The technological development in the launch vehicle design supported by the ever-increasing rate of success led to these launch vehicles being utilized for offering comparable services to the other nations that did not possess them (Jeffreys *et al.,* 1988; Gurfil *et al*., 2012).

On one side, there are countries intense to become self-reliant and thus attain a certain level of independence in this particular field; there are other nations whose commercial activities match the substantial part of the national activity (Gill *et al*., 2013; Bandyopadhyay *et al*., 2016).

Starting with the 1-stage R-7 rocket (called Semyorka) that put Sputnik-1 into orbit in the year 1957, Russia has designed numerous launch vehicles for several applications. Some prominent ones comprise the Vostok series, the Soyuz series, the Molniya series (Figure 1.32), the Proton series (Figure 1.33), the Energia series (Figure 1.34), and Zenit series (Royle *et al.,* 1988).

Energia is competent in placing the payload of 65–200 tons in the LEO.

Figure 1.32. *Molniya series.*

Source: Replicated by approval of © Mark Wade; http://www.astronautix.com/m/molniya-1.html.

Figure 1.33. *Proton series.*

Source: Image courtesy: NASA; https://bricksin.space/launch-vehicles/proton-series/.

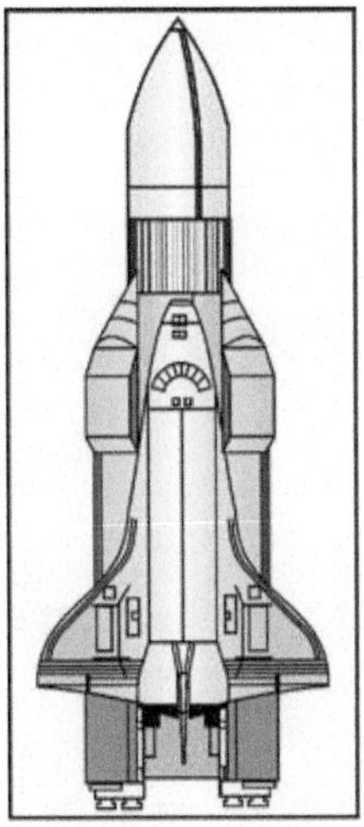

Figure 1.34. *Energia series.*

Source: Replicated by approval of © Mark Wade; https://catalogimages.wiley.com/images/db/pdf/9780470033357.excerpt.pdf.

Buran developed by Russia is one more reusable vehicle comparable in dimensions and design to the US Space Shuttle (Hearne et al., 1992; Baturkin, 2005). The key difference between the two is that the Buran does not have its specific propulsion system and is put into orbit with the help of the Energia launch vehicle (Figures 1.35–1.38).

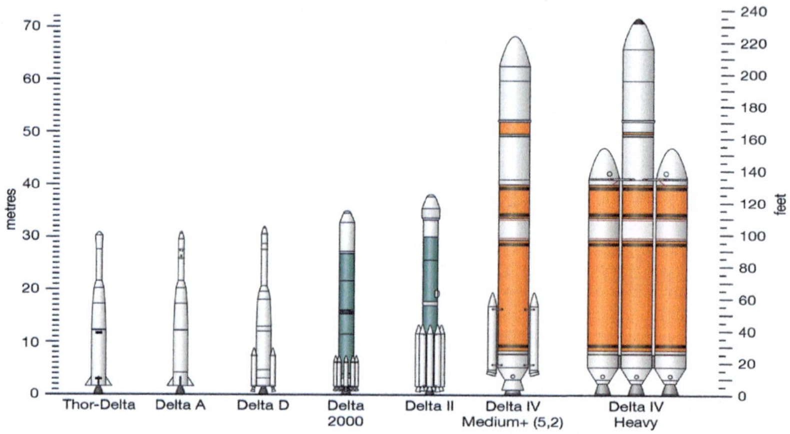

Figure 1.35. *Delta series.*

Source: https://www.britannica.com/technology/Delta-launch-vehicle.

Figure 1.36. *Titan series.*

Source: Image courtesy: NASA/JPL-Caltech; https://en.wikipedia.org/wiki/Titan_(rocket_family).

Figure 1.37. *Space shuttle.*

Source: Image courtesy: NASA; https://www.researchgate.net/figure/Space-Shuttle-Launch-Photo-courtesy-of-NASA_fig1_254545459.

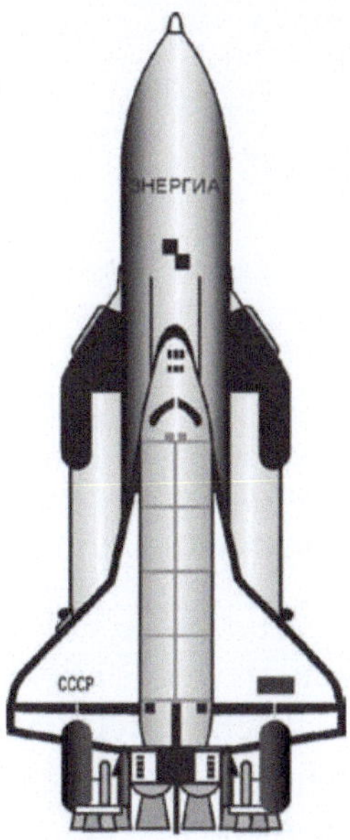

Figure 1.38. *Buran series.*

Source: https://www.buran.su/.

The launch vehicle named Ariane from the ESA (European Space Agency) has entered the 5th generation with its Ariane-5 heavy launch vehicle for satellites. Ariane-5 ECB (Improved Capability-B) has the capability of launching 12 tons to geostationary transfer orbit (GTO) (Figure 1.39) (Jarne and Lagoda, 1996).

Figure 1.39. *Ariane-5ECA.*

Source: Replicated by approval of © ESA-D. DUCROS; https://arstechnica.com/science/2018/07/as-the-spacex-steamroller-surges-european-rocket-industry-vows-to-resist/3/.

Long March from China, the geostationary satellite launch vehicle GSLV, and the polar satellite launch vehicle PSLV from India, and the series of H-2 from Japan are other functional launch vehicles. Noticeable launch vehicles, both reusable and expendable (Figures 1.40 and 1.41) (Bruford and Wayne, 1993; Goldstein and Pollock, 1997).

Figure 1.40. *Long march.*

Source: https://spaceflight101.com/chinese-long-march-3b-launches-ap-star-6c-communications-satellite/.

Figure 1.41. *GSLV.*

Source: Image courtesy: ISRO; https://economictimes.indiatimes.com/news/science/gslv-mark-iii-launch-why-isros-biggest-challenge-will-be-at-the-end-of-this-month/articleshow/58466879.cms.

1.12. FUTURE TRENDS

The technological developments in the area of satellites will usually be directed to reduce the size and cost of satellites along with enhancing the quality of services offered. The smaller satellites are developed as they can easily be launched with the help of smaller launchers, thus reducing the overall mission expense.

1.12.1. Communication Satellites

In the situation of communication satellites, main technologies comprise the development of the large-scale antenna having multiple beams to permit intensive re-utilization of frequencies, USAT terminals in order to substitute VSAT terminals, development of the signal processing algorithms to execute intelligent operations onboard the satellite comprising signal regeneration, overpowering the problem of signal fading because of rain, and permitting utilization of smaller antennas. Supple crosslink communication amongst the satellites will be established to permit good distribution of traffic amongst the satellites (Fairhurst *et al.*, 2008).

1.12.2. Weather Forecasting Satellites

These satellites in the future will carry with them advanced payloads comprising sounders, multispectral imagers, and scatterometers having a better resolution. These kinds of instruments will have nearly more than 1000 channels over the broad spectral range (Toyoshima, 2005).

The GOES-R satellite launched in 2015 carried various sophisticated instruments comprising the ABI (advanced baseline imager), SEISS (space environment *in-situ* suite), SIS (solar imaging suite), GLM (geostationary lightning mapper), and the Magnetometer. SEISS further includes two MIPS (magnetospheric particle sensors) (MPS-LO and MPS-HI), EHIS (energetic heavy ion sensor), and the SGPS (solar and galactic proton sensor). The SIS payload has the SUVI (solar ultraviolet imager), the XRS (solar X-ray sensor), and a EUVS (extreme ultraviolet sensor).

1.12.3. Earth Observation Satellites

For these kinds of satellites, technological improvements will incline to better resolution, upsurge in the observation area, and decrease in the access time. Plans for upcoming missions and instruments comprise completely new types of measurement technology, like hyperspectral sensors, lidars, cloud radars, and polarimetric sensors that will offer new insights into main parameters of atmospheric moisture and temperature, ocean salinity, and soil moisture. Various novel gravity field missions intended for more accurate determination of marine geoid will be put into in the future. These types of missions will focus on the disaster management and information of main Earth System processes-the carbon cycle, water cycle, cryosphere, the role of aerosols and clouds in worldwide climate change, and the sea-level rise (Cola *et al*., 2015).

1.12.4. Navigational Satellites

Satellite-centered navigation systems are further modernized to provide more reliable and accurate services.

The GPS is being updated to provide more reliable, accurate, and incorporated services to users. The first struggles in modernization started with the suspension of selective availability feature, to enhance the accuracy of civilian receivers. This aids in further enhancing accuracy by recompensing for the atmospheric delays and will guarantee greater navigation security.

The Block-IIF satellites have the third carrier signal at a frequency of 1176.45 MHz. The 1st Block-II F satellite was put into orbit in May 2010. Till August 2013, there were four functional Block-II F satellites in the GPS constellation. The GPS-III satellites series is at the stage of planning. These satellites will engage spot beams. The utilization of spot beams outcomes in augmented signal power, allowing the system to become more accurate and reliable, with system precision approaching a meter (m). Till August 2013, the Block III satellites were in the phase of production and deployment. The first Block-III satellite (Block IIIA-1) was launched in 2014.

The 1st phase or stage has been the experimental stage to experiment and verify the crucial technologies required for the Galileo navigation system to function in the MEO (medium earth orbit) environment. Two experimental satellites, Galileo in-orbit validation element-A (GIOVE-A) and GIOVE-B correspondingly launched on December 2005 and April 2008. The 2nd phase was meant to be the IOV (in-orbit validation) phase. The prime objective of this phase was validation of the system design utilizing the scaled-down constellation of just four satellites, which with the inadequate number of ground stations, is a minimum number required to provide timing and positioning data. The four satellites were put into orbit from pairs. The 1st pair of 2 satellites were launched in October 2011 and the 2nd pair was launched in October 2012. The 3rd phase fully accomplished the operational

Galileo system including 30 satellites (27 functional and 3 active spares), positioned in 3 circular MEO planes at a 23,222-kilometer altitude above the planet Earth (Labrador and Galace, 2005).

1.12.5. Military Satellites

The scope of use of the military satellites will increase further to offer the variety of services varying from communication services to collecting intelligence imagery data, from giving navigation information to giving timing data, and from weather forecasting data to initial warning applications. They have nowadays become an essential component of military planning of several developed as well as other countries. These types of satellites will be used to destroy the nuclear missiles in their initial phase inside the country that is trying to launch them (Labrador and Galace, 2005).

REFERENCES

1. Arnon, S., & Kopeika, N. S., (1997). Laser satellite communication network-vibration effect and possible solutions. *Proceedings of the IEEE, 85*(10), 1646–1661.
2. Bacsardi, L., (2013). On the way to quantum-based satellite communication. *IEEE Communications Magazine, 51*(8), 50–55.
3. Bandyopadhyay, S., Foust, R., Subramanian, G. P., Chung, S. J., & Hadaegh, F. Y., (2016). Review of formation flying and constellation missions using nanosatellites. *Journal of Spacecraft and Rockets, 53*(3), 567–578.
4. Baturkin, V., (2005). Micro-satellites thermal control—concepts and components. *Acta Astronautica, 56*(1, 2), 161–170.
5. Belward, A. S., & Skøien, J. O., (2015). Who launched what, when and why; trends in global land-cover observation capacity from civilian earth observation satellites. *ISPRS Journal of Photogrammetry and Remote Sensing, 103,* 115–128.
6. Bhasin, K. B., Bostian, C. W., Brandon, W. T., Evans, J. V., Helm, N. R., Mahle, C. E., & Townes, S. A., (1998). *Global Satellite Communications Technology and Systems, 1*, 1–44.
7. Block, B. A., Dewar, H., Farwell, C., & Prince, E. D., (1998). A new satellite technology for tracking the movements of Atlantic bluefin tuna. *Proceedings of the National Academy of Sciences, 95*(16), 9384–9389.
8. Bodroghkozy, A., (2016). The BBC and the black weekend: Broadcasting the Kennedy assassination and the birth of global television news. *The Sixties, 9*(2), 242–260.
9. Bruford, M. W., & Wayne, R. K., (1993). Microsatellites and their application to population genetic studies. *Current Opinion in Genetics and Development, 3*(6), 939–943.
10. Cola, T. D., Tarchi, D., & Vanelli-Coralli, A., (2015). Future trends in broadband satellite communications: Information centric networks and enabling technologies. *International Journal of Satellite Communications and Networking, 33*(5), 473–490.
11. Comparetto, G., & Ramirez, R., (1997). Trends in mobile satellite technology. *Computer, 30*(2), 44–52.
12. Elbert, B. R., (2008). *Introduction to Satellite Communication* (Vol. 1, pp. 1–22). Artech House.

13. Esper, J., Panetta, P. V., Ryschkewitsch, M., Wiscombe, W., & Neeck, S., (2000). NASA-GSFC nano-satellite technology for earth science missions. *Acta Astronautica, 46*(2–6), 287–296.
14. Evans, J. V., (2001). Communication satellite systems for high-speed internet access. *IEEE Antennas and Propagation Magazine, 43*(5), 11–22.
15. Fairhurst, G., Caviglione, L., & Collini-Nocker, B., (2008). FIRST: Future internet: A role for satellite technology. In: *2008 IEEE International Workshop on Satellite and Space Communications* (Vol. 1, pp. 160–164). IEEE.
16. Fernández, F. A., (2011). Inter-satellite ranging and inter-satellite communication links for enhancing GNSS satellite broadcast navigation data. *Advances in Space Research, 47*(5), 786–801.
17. Fugono, N., Hayashi, R., & Ishizawa, Y., (1980). ETS-II experiments. Part I: Japan's first geostationary satellite. *IEEE Transactions on Aerospace and Electronic Systems*, (5), 549–557.
18. Gan, Z. M., & Zhang, G. X., (2006). Current development of satellite communications technology. *Journal-China Institute of Communications, 27*(8), 2.
19. Gill, E., Sundaramoorthy, P., Bouwmeester, J., Zandbergen, B., & Reinhard, R., (2013). Formation flying within a constellation of nano-satellites: The QB50 mission. *Acta Astronautica, 82*(1), 110–117.
20. Goldstein, D. B., & Pollock, D. D., (1997). Launching microsatellites: A review of mutation processes and methods of phylogenetic inference. *Journal of Heredity, 88*(5), 335–342.
21. Golkar, A., & Cruz, I. L. I., (2015). The federated satellite systems paradigm: Concept and business case evaluation. *Acta Astronautica, 111*, 230–248.
22. Grami, A., (2006). A bent pipe v-band satellite system: Broadband mesh networking. In: *24th AIAA International Communications Satellite Systems* Conference (Vol. 1, pp. 5375).
23. Greenberg, J. S., (1979). Evaluating the economic impact of design alternatives on domestic communication satellite ventures. *Acta Astronautica, 6*(12), 1491–1513.
24. Gurfil, P., Herscovitz, J., & Pariente, M., (2012). *The SAMSON Project-Cluster Flight and Geolocation with Three Autonomous Nano-Satellites, 1*, 1–33.

25. Hall, D. K., (1988). Assessment of polar climate change using satellite technology. *Reviews of Geophysics, 26*(1), 26–39.
26. Hearne, C. M., Ghosh, S., & Todd, J. A., (1992). Microsatellites for linkage analysis of genetic traits. *Trends in Genetics, 8*(8), 288–294.
27. Heath, D. F., Hilsenrath, E., Krueger, A. J., Nordberg, W., Prabhakara, C., & Theon, J. S., (1972). *Observations of the Global Structure of the Stratosphere and Mesosphere with Sounding Rockets and with Remote Sensing Techniques from Satellites, 1*, 1–32.
28. Hnatushenko, V. V., Hnatushenko, V., Mozgovyi, D. K., & Vasiliev, V. V., (2016). Satellite technology of the forest fires effects monitoring. *Scientific Bulletin of National Mining University,* (1), 1–44.
29. Huang, J., Su, Y., Liu, W., & Wang, F., (2016). Adaptive modulation and coding techniques for global navigation satellite system inter-satellite communication based on the channel condition. *IET Communications, 10*(16), 2091–2095.
30. Iverson, L. R., Graham, R. L., & Cook, E. A., (1989). Applications of satellite remote sensing to forested ecosystems. *Landscape Ecology, 3*(2), 131–143.
31. Janson, S. W., (2020). I-1: A brief history of nanosatellites. *Nanosatellites: Space and Ground Technologies, Operations and Economics, 1*, 1–30.
32. Jarne, P., & Lagoda, P. J., (1996). Microsatellites, from molecules to populations and back. *Trends in Ecology and Evolution, 11*(10), 424–429.
33. Jeffreys, A. J., Wilson, V., Neumann, R., & Keyte, J., (1988). Amplification of human minisatellites by the polymerase chain reaction: Towards DNA fingerprinting of single cells. *Nucleic Acids Research, 16*(23), 10953–10971.
34. Kankaku, Y., Sagisaka, M., & Suzuki, S., (2014). PALSAR-2 launch and early orbit status. In: *2014 IEEE Geoscience and Remote Sensing Symposium* (Vol. 1, pp. 3410–3412). IEEE.
35. Karafantis, L., (2016). Telstar, the cold war, and the origins of global communications. *Vulcan, 4*(1), 112–134.
36. Kessler, D. J., & Cour-Palais, B. G., (1978). Collision frequency of artificial satellites: The creation of a debris belt. *Journal of Geophysical Research: Space Physics, 83*(A6), 2637–2646.

37. Kobayashi, T., (1995). Experimental evaluation of interference from domestic microwave ovens to 1.9-GHz digital radio transmission. In: *Proceedings of International Symposium on Electromagnetic Compatibility* (Vol. 1, pp. 293–296). IEEE.
38. Kogan, F. N., (1995). Droughts of the late 1980s in the United States as derived from NOAA polar-orbiting satellite data. *Bulletin of the American Meteorological Society, 76*(5), 655–668.
39. Koike, J., (1991). Fundamental questions concerning the contamination of other planets with terrestrial micro-organisms carried by space-probes. *The Journal of Space Technology and Science, 7*(2), 2_9–2_14.
40. Kourogiorgas, C. I., Lyras, N., Panagopoulos, A. D., Tarchi, D., Vanelli-Coralli, A., Ugolini, A., & Arapoglou, P. D., (2017). Capacity statistics evaluation for next generation broadband MEO satellite systems. *IEEE Transactions on Aerospace and Electronic Systems, 53*(5), 2344–2358.
41. Kramer, H. J., & Cracknell, A. P., (2008). An overview of small satellites in remote sensing. *International Journal of Remote Sensing, 29*(15), 4285–4337.
42. Kuang, L., Chen, X., Jiang, C., Zhang, H., & Wu, S., (2017). Radio resource management in future terrestrial-satellite communication networks. *IEEE Wireless Communications, 24*(5), 81–87.
43. Kudsia, C., Cameron, R., & Tang, W. C., (1992). Innovations in microwave filters and multiplexing networks for communications satellite systems. *IEEE Transactions on Microwave Theory and Techniques, 40*(6), 1133–1149.
44. Kumar, K., & Kumar, K. D., (2001). Feasibility of using satellites in non-equatorial 24-hour circular orbits for communications. *Acta Astronautica, 48*(5–12), 589–596.
45. Lambright, W. H., (1994). The political construction of space satellite technology. *Science, Technology, and Human Values, 19*(1), 47–69.
46. Li, T., Zhou, H., Luo, H., Xu, Q., & Ye, Y., (2016). Using SDN and NFV to implement satellite communication networks. In: *2016 International Conference on Networking and Network Applications (NaNA)* (Vol. 1, pp. 131–134). IEEE.
47. Livingston, S., & Belle, D. A. V., (2005). The effects of satellite technology on newsgathering from remote locations. *Political Communication, 22*(1), 45–62.

48. Lutz, E., Cygan, D., Dippold, M., Dolainsky, F., & Papke, W., (1991). The land mobile satellite communication channel-recording, statistics, and channel model. *IEEE Transactions on Vehicular Technology, 40*(2), 375–386.
49. Ma, L., Xie, P., Liu, D., & Wu, Y., (2019). Research on the influence of China's commercial spaceflight on the economic and social development of the regions along the belt and road. *New Space, 7*(4), 223–234.
50. Misra, D., Misra, D. K., & Tripathi, S. P., (2013). Satellite communication advancement, issues, challenges, and applications. *International Journal of Advanced Research in Computer and Communication Engineering, 2*(4), 1681–1686.
51. Muri, P., & McNair, J., (2012). A survey of communication sub-systems for intersatellite linked systems and CubeSat missions. *JCM, 7*(4), 290–308.
52. Nagata, T., Yoshino, T., Hirasawa, T., & Fukunishi, H., (1976). *Substorm Observations by Sounding Rockets and by Reception of Polar-Orbiting Satellite Data, 1, 1-20.*
53. Petrov, A. A., Davydov, V. V., Myazin, N. S., & Kaganovskiy, V. E., (2017). Rubidium atomic clock with improved metrological characteristics for satellite communication system. In: *Internet of Things, Smart Spaces, and Next Generation Networks and Systems* (Vol. 1, pp. 561–568). Springer, Cham.
54. Price, D. C., Enriquez, J. E., Brzycki, B., Croft, S., Czech, D., DeBoer, D., & Worden, S. P., (2020). The breakthrough listen search for intelligent life: Observations of 1327 nearby stars over 1.10–3.45 GHz. *The Astronomical Journal, 159*(3), 86.
55. Purchase, J., (1996). A conceptual design example of how a communications satellite could be produced for proposed' open network communications standards to provide interconnectivity between national communication satellite systems and the emerging North American broadband sy. In: *16th International Communications Satellite Systems Conference* (Vol. 1, p. 1067).
56. Radhakrishnan, R., Edmonson, W. W., Afghah, F., Rodriguez-Osorio, R. M., Pinto, F., & Burleigh, S. C., (2016). Survey of inter-satellite communication for small satellite systems: Physical layer to network layer view. *IEEE Communications Surveys and Tutorials, 18*(4), 2442–2473.

57. Rahmat-Samii, Y., & Densmore, A. C., (2014). Technology trends and challenges of antennas for satellite communication systems. *IEEE Transactions on Antennas and Propagation, 63*(4), 1191–1204.

58. Rahmat-Samii, Y., Manohar, V., & Kovitz, J. M., (2017). For satellites, think small, dream big: A review of recent antenna developments for CubeSats. *IEEE Antennas and Propagation Magazine, 59*(2), 22–30.

59. Renzetti, G., (2013). History of the attempts to measure orbital frame-dragging with artificial satellites. *Central European Journal of Physics, 11*(5), 531–544.

60. Riva, C., Capsoni, C., Luini, L., Luccini, M., Nebuloni, R., & Martellucci, A., (2014). The challenge of using the W band in satellite communication. *International Journal of Satellite Communications and Networking, 32*(3), 187–200.

61. Royle, N. J., Clarkson, R. E., Wong, Z., & Jeffreys, A. J., (1988). Clustering of hypervariable minisatellites in the proterminal regions of human autosomes. *Genomics, 3*(4), 352–360.

62. Sadek, M., & Aissa, S., (2012). Personal satellite communication: Technologies and challenges. *IEEE Wireless Communications, 19*(6), 28–35.

63. Sellers, P. J., Hall, F. G., Asrar, G., Strebel, D. E., & Murphy, R. E., (1992). An overview of the first international satellite land surface climatology project (ISLSCP) field experiment (FIFE). *Journal of Geophysical Research: Atmospheres, 97*(D17), 18345–18371.

64. Sharma, S. K., Chatzinotas, S., & Ottersten, B., (2013). Cognitive radio techniques for satellite communication systems. In: *2013 IEEE 78th Vehicular Technology Conference (VTC Fall)* (Vol. 1, pp. 1–5). IEEE.

65. Sweeting, M., (2018). Modern small satellites-changing the economics of space. *Proceedings of the IEEE, 106*(3), 343–361.

66. Takahashi, T., (2013). High energy astrophysics and fundamental physics missions in Japan. *Nuclear Physics B-Proceedings Supplements, 243*, 70–73.

67. Toyoshima, M., (2005). Trends in satellite communications and the role of optical free-space communications. *Journal of Optical Networking, 4*(6), 300–311.

68. Underwood, C., Pellegrino, S., Lappas, V. J., Bridges, C. P., & Baker, J., (2015). Using CubeSat/micro-satellite technology to demonstrate the autonomous assembly of a reconfigurable space telescope (AAReST).

Acta Astronautica, 114, 112–122.

69. Unwin, M., Jales, P., Tye, J., Gommenginger, C., Foti, G., & Rosello, J., (2016). Spaceborne GNSS-reflectometry on TechDemoSat-1: Early mission operations and exploitation. *IEEE Journal of Selected Topics in Applied Earth Observations and Remote Sensing, 9*(10), 4525–4539.
70. Wang, W., Tong, Y., Li, L., Lu, A. A., You, L., & Gao, X., (2019). Near-optimal timing and frequency offset estimation for 5G integrated LEO satellite communication system. *IEEE Access, 7*, 113298–113310.
71. Ward, J., (1993). *Design Implementation and in-Orbit Demonstration of a Store-and-Forward Digital Communication System for Low Earth Orbit Microsatellites* (Doctoral dissertation, Vol. 1, pp. 1–66).
72. Woods, T. N., Eparvier, F. G., Hock, R., Jones, A. R., Woodraska, D., Judge, D., & Viereck, R., (2010). Extreme ultraviolet variability experiment (EVE) on the solar dynamics observatory (SDO): Overview of science objectives, instrument design, data products, and model developments. *The Solar Dynamics Observatory, 1*, 115–143.
73. Wrenn, G. L., (1995). Conclusive evidence for internal dielectric charging anomalies on geosynchronous communications spacecraft. *Journal of Spacecraft and Rockets, 32*(3), 514–520.
74. Yajima, N., Izutsu, N., Imamura, T., & Abe, T., (2009). *Scientific Ballooning: Technology and Applications of Exploration Balloons Floating in the Stratosphere and the Atmospheres of Other Planets* (Vol. 112, pp. 1–32). Springer Science and Business Media.
75. Yan, X., An, K., Liang, T., Zheng, G., Ding, Z., Chatzinotas, S., & Liu, Y., (2019). The application of power-domain non-orthogonal multiple access in satellite communication networks. *IEEE Access, 7*, 63531–63539.
76. Yang, Y., Xu, Y., Li, J., & Yang, C., (2018). Progress and performance evaluation of BeiDou global navigation satellite system: Data analysis based on BDS-3 demonstration system. *Science China Earth Sciences, 61*(5), 614–624.
77. Zheng, G., Chatzinotas, S., & Ottersten, B., (2012). Generic optimization of linear precoding in multibeam satellite systems. *IEEE Transactions on Wireless Communications, 11*(6), 2308–2320.

Chapter

2 CLASSIFICATION OF KEY SATELLITE SYSTEMS

CONTENT

2.1. Introduction..44

2.2. Geosynchronous Earth Orbit Satellites..49

2.3. Medium Earth Orbit (Meo) Satellites..51

2.4. Low Earth Orbit (LEO) Satellites..53

2.5. Highly Elliptical Orbit (HEO), Basics..56

2.6. Polar Orbit and Sun-Synchronous Orbit (SSO)..58

2.7. Transfer Orbits and Geostationary Transfer Orbit (GTO)..59

2.8. Lagrange Points..60

2.9. Narrowband Leo, Meo, and Geo Systems..60

References

2.1. INTRODUCTION

A satellite communication system is distinguished through its inherent broadcast capability, global coverage, the capability to support flexibility, and bandwidth-on-demand mobility. Taking into description this developing communication role of the satellite systems, this chapter gives an outline of major narrowband MEO, LEO, and GEO satellite systems. It provides some typical commercial instances of such satellite systems and debates their basic features comprising of utilized frequency bands, supported terminals and applications, and critical performance problems (Tjelta *et al.,* 2001; Panagiotarakis *et al*., 2002). Numerous wireless and wired technologies are contending for providing high-bandwidth access. Speed, cost, flexibility, and time-to-market are key factors affecting the development of these technologies. Digital Subscriber Loops (xDSL) and cable modems are the most encouraging wired broadband access structures. MMDS systems function at frequencies lesser than 5 GHz with cell radius (coverage areas) up to 40 km. LMDS systems function at higher frequencies, where huge portions of the spectrum are still free. In this situation, the coverage area is realized with slighter cells (normally up to a radius of 5 km). The extension of this radius generally needs repeaters to be placed in a line of sight (LOS) configuration (Bem et al., 2000; Hu and Li, 2001).

Moving to the satellite systems, their communication role is not to contend with the land-based fixed, mobile, or wireless communications systems earlier mentioned; however, to complement them both in a "service complement" sense (satellite delivery is more suitable and cost-efficient for multicast /broadcast kind of services) and in a "geographical" sense (where it is economically impractical or impossible for the terrestrial structures to provide services coverage).

2.1.1. What Is an Orbit?

An orbit is a curved route that an object in space (like a moon, planet, star, spacecraft, or asteroid) takes about another object because of gravity.

Objects of the same mass orbit each other with no object at the center, while small objects orbit around larger objects. In our Solar System, the Moon orbits Earth, and Earth orbits the Sun; however, that does not mean the greater object remains entirely still. Due to gravity, Earth is pulled marginally from its center by the Moon (that is why tides form in our oceans) and our Sun is pulled marginally from its center by Earth and other planets (Lim *et al*., 2005; Gupta, 2011).

With the Sun being so much bigger than these little bits of gas and dust, its gravity enticed these bits into orbit around it, shaping the cloud into a type of ring around the Sun (Yizhou *et al*., 2015).

Ultimately, these particles initiated to settle and bunch together (or 'coalesce'), growing ever bigger like rolling snowballs unless they formed what we currently see as moons, asteroids, and planets. The reality that the planets were all made together this way is why all the planets have orbits around the Sun in a similar direction, in approximately the same plane (Figure 2.1) (Luu and Sadler, 2001).

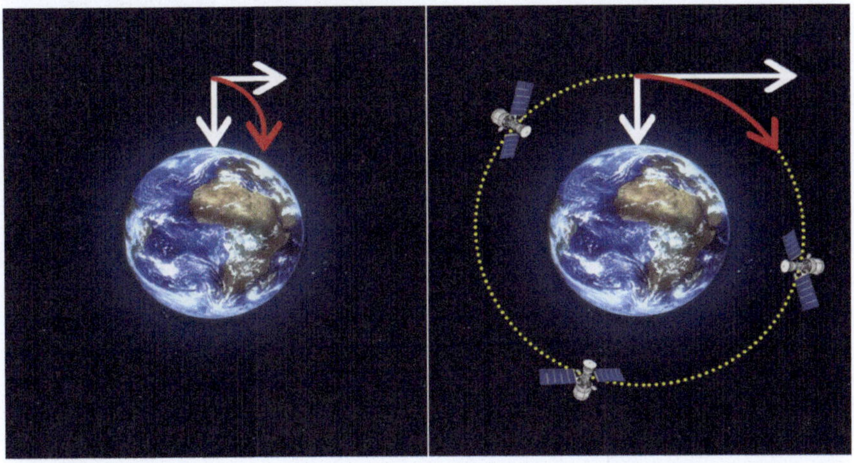

Figure 2.1. *A satellite reaching the orbit.*

Source: https://www.esa.int/Enabling_Support/Space_Transportation/Types_of_orbits.

When rockets takeoff our satellites, they place them into orbit in space. There, gravity retains the satellite on its needed orbit in the same way that gravity retains the Moon in orbit around Earth. While it is your throw that provides the ball its early speed, it is gravity alone that keeps the ball moving to the ground once you let go (Pratt *et al.,* 1999; Schöne *et al*., 2011).

In the same fashion, a satellite is placed into orbit by being positioned 100 or 1000 km above Earth's surface (as if in a very giant tower) and then provided a 'push' by the rocket's engines to make it initiate on its orbit (Alagoz and Gur, 2011; Sabetghadam *et al.,* 2021).

As revealed in the figure, the change is that throwing something would make it fall on a curved path to the ground; however, actually energetic throw would mean that the ground initiates to curve away earlier your object touches the ground. Congratulations! You have reached orbit (Corcoran *et al.*, 2013).

In space, there is no air and thus no air resistance, so gravity takes the satellite orbit around Earth with nearly no further support. Placing satellites into orbit allows us to utilize technologies for navigation, astronomy observations, weather forecast, and telecommunication (Figure 2.2) (Kumar *et al.*, 2018).

Figure 2.2. *Artist's image of Europe's launcher family.*

Source: https://www.esa.int/ESA_Multimedia/Images/2018/10/Artist_s_view_of_Europe_s_launcher_family3.

2.1.2. Launch to Orbit

Europe's family of rockets function from Europe's Spaceport in Kourou, French Guiana. On every mission, a rocket positioned one or more satellites onto their discrete orbits (Logue and Pelton, 2019).

A high altitude or a heavy payload orbit needs more power to contest Earth's gravity than a lighter payload at a lesser altitude (Tronc *et al.*, 2014; Araguz *et al.*, 2018).

Ariane 5 is Europe's most influential launch vehicle, able of lifting one, two, or multiple satellites into their obligatory orbits. Relying on which orbit Ariane 5 is going to, it can launch amongst nearly 10 to 20 tons into space-that is 10,000 to 20,000 kg, which is around the weight of a city bus. Vega is smaller than Ariane 5, able to launch approximately 1.5 tons at a time, creating it a perfect launch vehicle for numerous scientific and Earth observation missions. Both Vega and Ariane 5 could deploy many satellites at a time (Key *et al.*, 2001; Wilkinson, 2005).

ESA's next generation of rockets comprises Vega-C and Ariane 6. These rockets would be more flexible and would extend what Europe is able of getting into orbit, and would be able to deliver payloads to numerous diverse orbits in a single flight like a bus with several stops (Colino, 1977; Goldhirsh, 1992).

2.1.3. Types of Orbit

Upon launch, a spacecraft or satellite is mostly often positioned in one of the numerous particular orbits around Earth or it might be directed on an interplanetary journey, meaning that it does not orbit Earth anymore, however instead orbits the Sun till it reaches its final destination, like Mars or Jupiter (Tomiyasu, 1978; Yan *et al.,* 2018).

Numerous factors select which orbit would be finest for a satellite to utilize, relying on what the satellite is designed to attain:

1. Low earth orbit (LEO);
2. Medium earth orbit (MEO);
3. Geostationary orbit (GEO);
4. Transfer orbits and geostationary transfer orbit (GTO);
5. Lagrange points (L-points);
6. Polar orbit and sun-synchronous orbit (SSO).

2.1.4. An Overview of Major Satellite Systems

The satellite should be appropriately positioned in the corresponding orbit after exiting it in space. It rotates in a specific way and attends its purpose for military, scientific, or commercial. The satellites existing in those orbits are termed Earth Orbit Satellites (Sohier *et al.*, 2014, 2015).

We should select an orbit properly for a satellite founded on the requirement. For instance, if the satellite is positioned in lower orbit, then it takes less time to travel around the Earth, and there would be a better resolution in an onboard camera. Likewise, if the satellite is positioned in a higher orbit, then it takes more time to travel around the Earth, and it covers additional Earth's surface at one time (Yang et al., 2016).

Following are the three significant kinds of Earth Orbit satellites (Sarigul-Klijn *et al.,* 2005; Oh *et al.,* 2013):

1. Geosynchronous earth orbit satellites (GEO);
2. Low earth orbit satellites (LEO); and
3. Medium earth orbit satellites (MEO).

Now, let us debate about each kind of earth orbit satellite one by one.

Figure 2.3 shows the paths of LEO, MEO, and GEO.

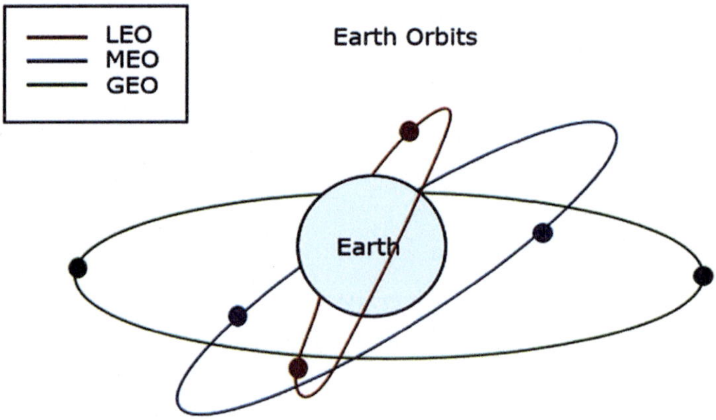

Figure 2.3. *Paths of GEO, MEO, and LEO.*

Source: https://www.tutorialspoint.com/satellite_communication/satellite_communication_earth_orbit_satellites.htm.

2.1.5. Orbital Slots

Here, a question might arise that with more than 200 satellites that are in geosynchronous orbit, how do we retain them from running into each other or from endeavoring to utilize the same place in space?

To reply to this question (issue), international regulatory bodies like the ITU (International Telecommunications Union) and national government organizations like the FCC (Federal Communications Commission) elect the locations on the geosynchronous orbit, where the communications satellites could be positioned (Richardson, 1972; Jäggi et al., 2007).

The ITU and FCC have progressively decreased the obligatory spacing down to merely 2° for ITU and C-band satellites because of the enormous demand for orbital slots (Xu et al., 2011).

2.2. GEOSYNCHRONOUS EARTH ORBIT SATELLITES

A GEO (Geo-synchronous Earth Orbit) Satellite is one, which is positioned at an altitude of 22,300 miles above the Earth. This orbit is coordinated with a sidereal day (for example, 23 hours 56 minutes). This orbit could be sloped at the poles of the Earth. However, it seems stationary when noticed from the Earth. These satellites are utilized for satellite Television (Hikami et al., 1980; Ferguson *et al.,* 2015). A similar geosynchronous orbit, if it is spherical and in the plane of the equator, then is termed as geostationary orbit (GEO). These Satellites are positioned at 35,900 km (same as geosynchronous) above the Earth's Equator, and they are kept on spinning concerning the Earth's direction (west to east). Therefore, these satellites are deliberated as stationary for the Earth as these are synchronous with the Earth's spin (Kirillov, 1999; Oltrogge *et al.,* 2018). The benefit of the GEO is that not required to track the antennas to locate the position of satellites. GEO (Geostationary Earth Orbit) Satellites are utilized for satellite TV, satellite radio, weather forecasting, and other kinds of global communications (Byers and Dankanich, 2008; Cognion, 2014). Figure 2.4 displays the difference between geostationary and geosynchronous orbits. The axis of rotation shows the movement of Earth.

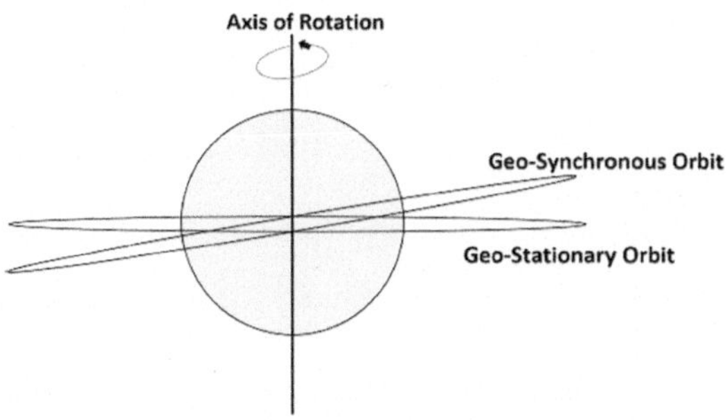

Figure 2.4. *Diagram of geosynchronous and geostationary orbits.*

Source: https://www.scienceabc.com/nature/universe/what-is-a-geosynchronous-satellite-and-how-is-it-different-from-a-geostationary-satellite.html.

Note: Each Geostationary orbit is a Geosynchronous orbit. However, the converse need not be factual.

The first main geosynchronous satellite project was the Defense Department's ADVENT communications satellite. It was 3-axis stabilized despite spinning. At 500 to 1000 pounds, it could merely be launched through the ATLAS-CENTAUR launch vehicle. ADVENT not ever flew, mainly because the CENTAUR stage was not completely reliable till 1968, however also because of issues with the satellite. When the program was negated in 1962, it was perceived as the death knell for geosynchronous satellites, the ATLAS-CENTAUR, three-axis stabilization, and complicated communications satellites normally. Geosynchronous satellites became a fact in 1963 and became the only option in 1965. The other ADVENT features also became a general place in the years to follow (Friesen et al., 1992).

In the early 1960s, transformed ICBMs (intercontinental ballistic missiles) and IRBMs (intermediate-range ballistic missiles) were utilized as launch vehicles. These all had general issues: they were designed to send an object to the Earth's surface, not to position an object in orbit. The DELTA launch vehicles, which positioned all the previous communications satellites in orbit, were THOR IRBMs that utilized the vanguard upper stage to give this delta-Vee. ATLAS-CENTAUR became dependable in 1968, and the 4th generation of INTELSAT satellites utilized this launch vehicle. The 5th-generation utilized ATLAS CENTAUR and a novel launch vehicle, the European ARIANE. Subsequently that time other entries, comprising the Russian PROTON launch vehicle and the Chinese LONG MARCH had come into the market. All are able of launching satellites almost 30 times the weight of EARLY BIRD (Alfriend *et al.,* 2006).

In the mid-1970s numerous satellites were made using 3-axis stabilization. They were more complicated than the spinners, however, they give more despond surfaces to support antennas, and they made it probable to deploy very huge solar arrays. Perhaps the surest sign of the accomplishment of this form of stabilization was the switch of Hughes, closely known with spinning satellites, to this kind of stabilization in the early 1990s (Hudson and Kolosa, 2020).

The newest products from the manufacturers of SYNCOM look relatively the same as the discredited ADVENT design of the late 1950s (Figure 2.5).

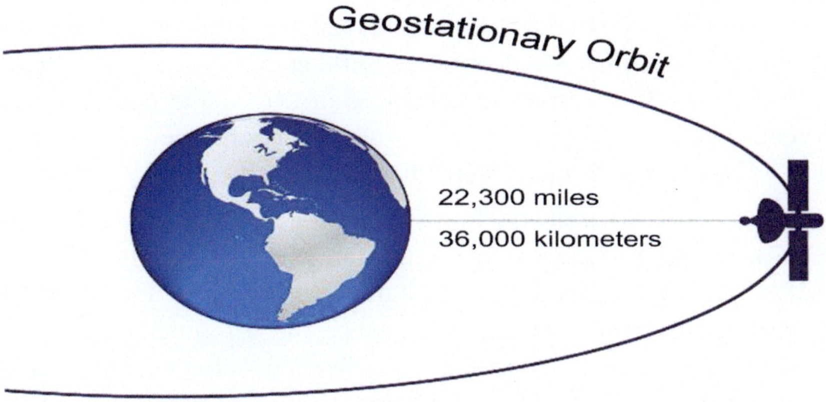

Figure 2.5. *Drawing of geostationary orbit.*

Source: https://wiki.pathfinderdigital.com/wiki/geo/.

A satellite in a GEO seems to be in a static position to an earth-based viewer. A geostationary satellite rotates around the Earth at a continual speed once per day over the equator.

The GEO is beneficial for communications applications due to ground-built antennas, which must be focused toward the satellite, could operate effectively without the requirement for costly equipment to track the satellite's motion.

2.3. MEDIUM EARTH ORBIT (MEO) SATELLITES

Medium earth orbit (MEO) satellites would orbit at distances of around 8000 miles from the Earth's surface. Signals transferred from an MEO satellite travel a shorter distance. This demonstrates that lightweight and smaller receiving terminals could be utilized at the receiving end (Luo *et al.*, 2019).

Transmission delay could be defined as the time a signal takes to travel up to a satellite and back down to a receiving station. In this situation, there is less transmission delay. For real-time communications, the little the transmission delay, the improved would be the communication system. As an instance, if a GEO satellite needs 0.25 seconds for a round trip, then the MEO satellite needs less than 0.1 seconds to finish a similar trip. These satellites are utilized for high-speed telephone signals. Ten or more MEO satellites are needed to cover the complete Earth (Tomiyasu, 1978).

An MEO satellite is a satellite that orbits the Earth in between LEOS, which orbit the Earth at a distance of around 200–930 miles (321.87–1496.69 km) and those satellites which orbit the Earth at GEO, around 22,300 miles (35,888.71 km) above the Earth. Each kind of satellite could provide a different kind of coverage for wireless and communications devices. Like LEOs, MEOs do not sustain a static distance from the Earth.

Any satellite that orbits the Earth amongst around 1000 to 22,000 miles (1609.34–35,405.57 km) above Earth is an MEO. Normally the orbit of an MEO satellite is around 10,000 miles (16,093.44 km) above the Earth. In numerous patterns, these satellites make the trip around the Earth in somewhere from 2 to 12 hours, which delivers better coverage to broader areas than that delivered by LEOs (Verstraete *et al.,* 2018).

In 1962, the initial communications satellite, Telstar, was launched. It was an MEO satellite designed to facilitate high-speed telephone signals, however, scientists shortly learned what some of the problematic features were of a single MEO in space. It only delivered transatlantic telephone signals for 20 min of every approximately 2.5 hours of orbit. It was obvious that multiple MEOs were required to be utilized to provide constant coverage.

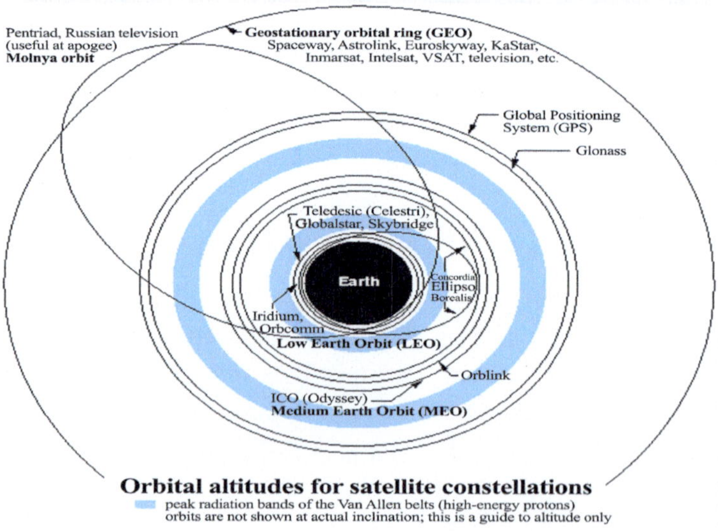

Figure 2.6. *Comparison of medium earth orbital and low earth orbital.*

Source: http://personal.ee.surrey.ac.uk/Personal/L.Wood/constellations/tables/overview.html.

Since then, several companies have launched both MEOs and LEOs. You required around two dozen LEOs to provide constant coverage and fewer MEOs. A MEOs might have a diversity of different orbits, comprising elliptical ones, and might provide better whole coverage of satellite communications, if adequate of them are in position and the orbit is swift. The coverage of Earth is termed a footprint, and MEOs normally are capable of forming a larger footprint due to their different orbital patterns, and since they are higher than LEOs.

Currently, the MEOs are most commonly utilized in navigation systems around the globe. These comprise GPS (global positioning system), and the Russian GLONASS. A planned MEO navigation system for the EU (European Union) termed Galileo is expected to initiate operations in 2013 (Figure 2.6).

2.4. LOW EARTH ORBIT (LEO) SATELLITES

Low earth orbit (LEO) satellites are primarily classified into three categories. Those are Mega LEOs, big LEOs, and little LEOs. LEOs will orbit at a distance of around 500 to 1000 miles above the Earth's surface. This comparatively short distance decreases transmission delay to merely 0.05 seconds. This further decreases the requirement for bulky and sensitive getting equipment. Twenty or more LEOS are needed to cover the whole Earth. Mega-LEOs function in the 20–30 GHz range, Big LEOs would function in the 2 GHz or above range, and Little LEOs would function in the 800 MHz (0.8 GHz) range (Huang et al., 2010; Blumenthal, 2013).

In February 1976, COMSAT launched a new type of satellite, MARISAT, to deliver mobile services to the US Navy and other maritime clients. In the initial 1980s, the Europeans launched the MARECS series to give similar services. In 1979 the UN International Maritime Organization backed the founding of the INMARSAT (International Maritime Satellite Organization) in a manner same to INTELSAT. INMARSAT initially leased the MARECS and MARISAT satellite transponders, however in Oct 1990, it launched its first own satellites, INMARSAT II F-1. The third generation, INMARSAT III, had previously been launched (Lücking *et al.*, 2012).

An aeronautical satellite was suggested in the mid-1970s. A contract was given to General Electric to build the satellite; however, it was void. INMARSAT now gives this service. Although INMARSAT was firstly considered as a method of giving telephone service and traffic-monitoring facilities on ships at sea, it had delivered much more. The journalist with a briefcase phone had been abundant for a certain time, however, the Gulf War brought this technology to the general eye (Skoulidou et al., 2019). In the subsequent year, the first MSAT satellite, in which TMI (Canada) and AMSC (U.S) collaborate, would be launched, giving mobile telephone service through satellite to all of North America (Rossi, 2008).

When EARLY BIRD was launched in 1965, the satellite provided nearly 10 times the capability of the submarine telephone cables for nearly 1/10th the price. TAT8 was the initial fiber-optic cable placed across the Atlantic. Satellites are still cheap with cable for point-to-point communications; however, the future benefit might lie with fiber-optic cable. Satellites still had two benefits over cable: they are more reliable and they could be utilized point-to-multi-point (broadcasting).

Long-distance calls need some other technology; however, this could be either fiber-optic cable or satellites (Ryden *et al.,* 2015).

Cellular telephony has provided us a novel technological "system"-the PCS (personal communications system). In the completely developed PCS, the individual would take his telephone with him. This telephone could be utilized for data or voice and would be operational anywhere. Numerous companies have committed themselves to give a form of this system utilizing satellites in LEO. The formerly "low-orbit" satellites were in oval orbits that took them by the lower van Allen radiation belt. The new structures would be in orbits at around 500 miles, below the belt (Lyras et al., 2019; Trishchenko et al., 2019).

The most striving of these LEO systems are Iridium, backed by Motorola. Iridium proposed to launch 66 satellites into a polar orbit at altitudes of around 400 miles. Each of the six orbital planes, divided by 30° around the equator, would comprise 11 satellites. Iridium originally proposed to have 77 satellites later its name. Element 66 has a little pleasant name Dysprosium. Iridium assumes to be giving communications services to handheld telephones in 1998. The entire cost of the Iridium system is well besides 3 billion dollars (Corazza and Vatalaro, 1994).

In addition to the "Big LEOS" like Globalstar and Iridium, there are numerous "little LEOs." These companies plan to provide more limited services, normally radio and data determination. Generally, of these is ORBCOM which had previously launched an experimental satellite and assumes to offer restricted service very soon (Figure 2.7).

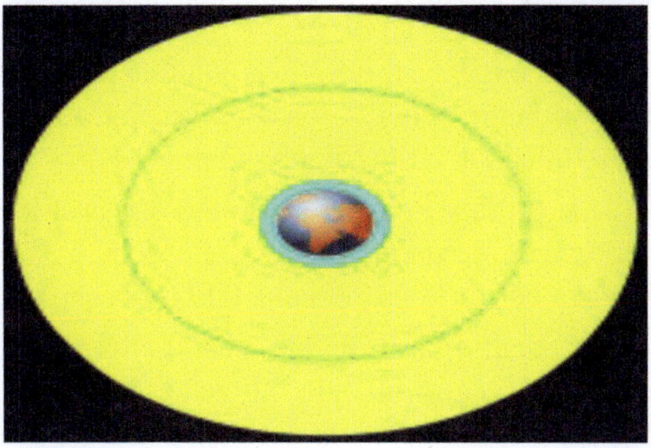

Figure 2.7. *Drawing of low earth orbitals.*

Source: https://enacademic.com/dic.nsf/enwiki/30429.

An LEO usually is a circular orbit around 400 kilometers above the Earth's surface and, congruently, a period (time to rotate around the Earth) of around 90 minutes. Due to their low altitude, these satellites are only observable from within a radius of approximately 1000 kilometers from the sub-satellite point. So even for local applications, a great number of satellites are required if the mission needs continuous connectivity.

LEOS are less costly to launch into orbit than geostationary satellites and, because of proximity to the ground, do not need as high signal strength (Recall that signal strength drops off as the square of the distance from the source, hence the effect is dramatic). Also, there are significant differences in the ground and onboard equipment required to support the two kinds of missions.

A group of satellites functioning in concert is recognized as a satellite constellation. Two such constellations, proposed to deliver satellite phone services, mainly to remote areas, are the Globalstar and Iridium systems. The Iridium system has 66 satellites. Additional LEOS constellation identified as Teledesic, with assistance from Microsoft entrepreneur Paul Allen, was to have above 840 satellites. This was future mounted back to 288 and finally ended up merely launching a single test satellite.

It is also probable to provide discontinuous coverage utilizing a LEOS able of storing data received through passing over one portion of Earth and transmitting it future though passing over another portion.

2.5. HIGHLY ELLIPTICAL ORBIT (HEO), BASICS

Though circular orbits might be the obvious solution for numerous satellites, elliptical orbits have many benefits for certain applications. The elliptical orbit is frequently called the HEO (highly elliptical orbit).

As a consequence of this, numerous satellites are positioned in elliptical orbits, particularly where certain attributes are needed. For instance, it does not need the orbits to be equatorial like the GEO.

The satellite elliptical orbit provides several coverage choices that are not accessible when circular orbits are utilized. As the name indicates, an elliptical orbit, or as it is more usually known, the HEO. HEO follows the curve of an ellipse. However, one of the main features of an elliptical orbit is that the satellite in an elliptical orbit around Earth moves much quicker when it is near to Earth than when it is further away (Ahedo and Sanmartin, 2002; De Weck *et al.,* 2004).

For any ellipse, there are 2 focal points, and one of them is the geo-center of the Earth. An additional feature of an elliptical orbit is that there are two further main points. The point where it is nearest to the Earth is termed as the perigee; this is the point where the satellite moves at its fastest (Figure 2.8) (Pratt et al., 1999).

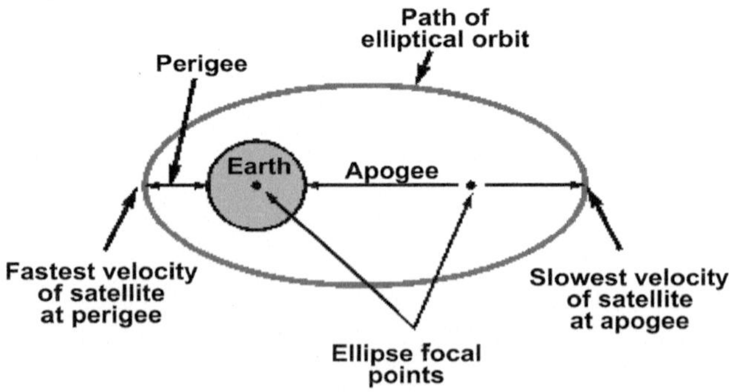

Figure 2.8. *Highly elliptical satellite orbit (HEO).*

Source: https://www.electronics-notes.com/articles/satellites/basic-concepts/highly-elliptical-orbit-heo.php.

If the satellite orbit is much elliptical, the satellite would spend the maximum of its time near apogee, where it moves very gradually. This means that the satellite could be in view above its operational area for a maximum of the time, and dropping out of view when the satellite comes nearer to the Earth and moves over the blind side of the Earth.

The plane of a satellite orbit is also significant. Some might orbit around the equator; however, others might have different orbits. The angle of inclination of a satellite orbit is revealed below. It is the angle amongst a line perpendicular to the plane of the orbit and a line passing through the poles. This shows that an orbit directly upper of the equator would incline 0° (or 180°), and one moving over the poles would have an angle of 90° (Figure 2.9) (Chang *et al*., 1998).

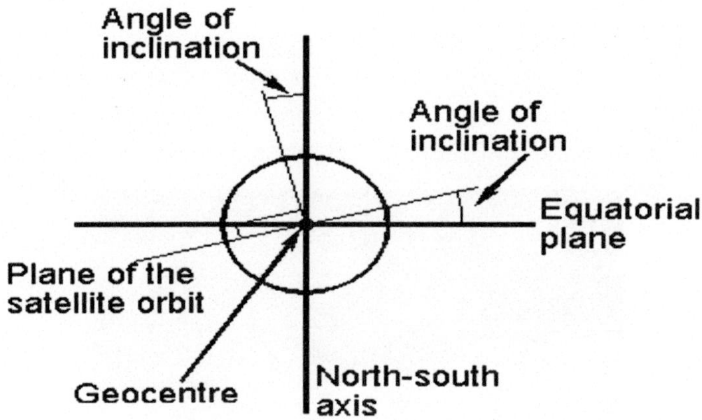

Figure 2.9. *The angle of inclination of a satellite orbit.*

Source: https://www.electronics-notes.com/articles/satellites/basic-concepts/satellite-orbits-types-definitions.php.

Those orbits overhead the equator is usually termed equatorial obits, while those above the poles are termed polar orbits. The highly elliptical satellite orbit could be utilized to provide a covering over any portion of the globe. As a consequence, its capability to provide polar coverage and high latitude, countries like Russia, which require coverage over polar and nearby polar areas, make significant utilization of HEO. With two satellites in any orbit, they are capable to provide constant coverage. The main drawback is that the satellite location from a point on the Earth does not remain identical (Del Portillo *et al.,* 2019).

2.6. POLAR ORBIT AND SUN-SYNCHRONOUS ORBIT (SSO)

Satellites in polar orbits typically travel past Earth from north to south instead of from west to east, moving roughly over Earth's poles. Satellites in a polar orbit do not have to pass the North and South Pole exactly; even an abnormality within 20 to 30° is yet categorized as a polar orbit. Polar orbits are a kind of LEO, as they are at short altitudes amongst 200 to 1000 km.

SSO (Sun-synchronous orbit) is a specific kind of polar orbit. Satellites in SSO, moving over the Polar Regions, are synchronous with the Sun. This reveals they are synchronized to constantly be in a similar 'fixed' position comparative to the Sun. This means that the satellite would always perceive a point on the Earth as if continually at a similar time of the day, which serves several applications; for instance, it means that scientists and those who utilize the satellite images could compare how somewhere alterations over time (Trishchenko and Garand, 2012).

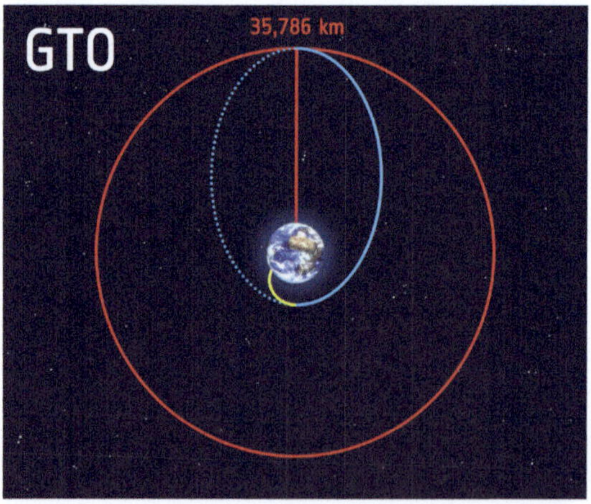

Figure 2.10. *Launch and rising to space (yellow line) develops the GTO (geostationary transfer orbit) (blue line) when the rocket discharges the satellite in space on a path to geostationary orbit (red line).*

Source: https://www.esa.int/ESA_Multimedia/Images/2020/03/Geostationary_transfer_orbit.

This is if you want to observe an area by taking a series of images of a specific place across several days, weeks, months, or even years, then it would not be very supportive to associate somewhere at midnight, and then at midday, you need to capture every picture as similarly as the earlier picture as possible. Therefore, scientists utilize image series like these to examine how weather patterns appear, to assist predict storms or weather; when monitoring emergencies like flooding or forest fires; or to gather data on long-term issues like rising sea levels or deforestation (Colombo, 2016).

Frequently, satellites in SSO are synchronized so that they are in constant dawn or dusk; this is because through constantly riding a sunrise or sunset, they would not ever have the Sun at an angle where the Earth shades them. A satellite in an SSO would typically be at an altitude of 600 to 800 km. At 800 km, it would be traveling at a speed of approximately 7.5 km/second (Figure 2.10) (Kahr *et al.*, 2014).

2.7. TRANSFER ORBITS AND GEOSTATIONARY TRANSFER ORBIT (GTO)

Transfer orbits are a special type of orbit utilized to get from one orbit to another. When satellites are launched from Earth and moved to space with launch vehicles like Ariane 5, the satellites are not constantly placed rightly on their last orbit. (Weeden and Shortt, 2008; Efimov *et al.*, 2018).

This permits a satellite to reach, for instance, a high-altitude orbit like GEO without really requiring the launch vehicle to move to this altitude, which would need more effort; this is similar to taking a shortcut.

Orbits had diverse eccentricities, a measure of how elliptical (squashed) or circular (round) an orbit is. In a flawlessly round orbit, the satellite is permanently at a similar distance from the Earth's surface on a highly eccentric orbit like this, the satellite could quickly move from being very distant to very near Earth's surface, relying on where the satellite is in orbit. In transfer orbits, the payload utilizes engines to move from an orbit of one eccentricity to another, which places it on track to lower or higher orbits.

After takeoff, a launch vehicle makes its path to space after a path shown through the yellow line, in Figure 2.10. At the target endpoint, the rocket discharges the payload, which sets it off on an elliptical orbit, showing the blue line which directs the payload farther away from Earth. The point furthest away from the Earth on the blue elliptical orbit is termed the apogee, and the point nearest is termed the perigee (Figure 2.11) (Paek *et al.,* 2020).

Figure 2.11. *The ESA telescope Gaia orbits around an L-point. The point is accurately behind earth, so at this point, Gaia would be in earth's shadow and incapable to receive the sunlight required to power its solar panels. Every few years, Gaia usages its motors to regulate its position to sustain this orbit.*

Source: https://www.esa.int/Enabling_Support/Space_Transportation/Types_of_orbits.

2.8. LAGRANGE POINTS

For numerous spacecraft being placed in orbit, being too near to Earth could be disrupting their mission, however, at more distant orbits like GEO. Photographing dark space through a telescope after our glowing Earth would be as disheartened as trying to getting pictures of stars from Earth in expansive daylight (Wang and Gurfil, 2016; Wang *et al.,* 2020). L-points or Lagrange points permit orbits that are much distant away (above a million kilometers) and do not orbit Earth directly. If a spacecraft was propelled to other points in space very detached from Earth, they would fall into an orbit around the Sun, and those spacecraft would presently end up far from Earth, making communication tough. As a spare for, spacecraft propelled to these special L-points remain fixed, and near to Earth with marginal effort without going into a diverse orbit (Morand et al., 2012).

2.9. NARROWBAND LEO, MEO, AND GEO SYSTEMS

2.9.1. Iridium

The Iridium system is founded on the concept of 66 orbiting satellites in 6 polar planes giving 11 satellites in every orbit. The system plan is technologically inspiring. Iridium employs OBP, ISLs, and satellite-founded switching. These technologies are not still effectively tested as mainly of other prospective MEO or LEO-founded mobile satellite systems make utilization of bent pipe satellites. Each satellite is connected to two others in its plane and two in contiguous planes (Messenger *et al.*, 2014).

2.9.2. Globalstar

The Globalstar space section comprises 48 spacecraft in LEO, flying in 8 orbital planes persuaded by 52° at an altitude of 1414 km overhead ground. Globalstar aims to deliver services on Earth from 70° south latitude to 70° north latitude. This inclination had been selected to produce the highest satellite density at the latitudes of the maximum population density. The spacecraft plan life is 7.5 and up to 10 years (Baoyin and McInnes, 2006).

The Globalstar payload contains transparent passive transponders minus switching or OBP. The user segment contains mobile, fixed, and personal dual-mode cellular/Globalstar terminals. The satellite broadcast is asymmetrical.

Relying on Globalstar service provides infrastructure and policy extra services might comprise data and facsimile transmission. The usual sustained rate for voice is 2.4 kbit/s; however, the extremely supported data rate is 9.6 kbit/s (Broucke, 1979).

2.9.3. Thuraya

The Thuraya mobile satellite system would be the result of a regional project made by Boeing Satellite Systems. The project comprises the manufacture of 2 high-power geosynchronous satellites, the launch of the initial satellite, the manufacturing and the fitting of the ground network equipment, and the manufacturing of approximately a quarter of a million mobile handsets. After the project, Thuraya would cover an area of almost 99 countries that are occupied by 2.3 billion people.

Thuraya's initial satellite was launched in October 2000, and commercial services started in a gradual rollout in several nations in 2001. Thuraya's system had been adapted for effective operation in both GSM and satellite environments. Thuraya's satellites had been particularly designed to attain a network capacity of around 13,750 telephone channels (Gineste et al., 2017).

Thuraya offers vehicular, fixed, and hand-held terminals to cater to the requirements of its subscribers. The services comprise fax (ITU-T G3 at

2.4, 4.8 and 9.6 kbps), messaging (GSM short message service), location determination (within 100 meters precision utilizing the GPS), voice (GSM quality), and data (at 2.4, 4.8, and 9.6 kbps), Thuraya would enhance its mobile satellite service portfolio to give broadband services through Inmarsat.

2.9.4. ACeS

The AceS (Asian cellular system) is a GEO regional system relatively the same in market location and philosophy with Thuraya that covers India, China, South East Asia, and Australia. Its initial satellite was launched effectively in February 2000, and ACeS provides services in the Philippines and Indonesia with a plan to do the similar in nearby future in other Asian countries (Nicholson and Gerstein, 2000).

2.9.5. Inmarsat

The GEO satellite systems functioned by the Inmarsat (INMARSAT) cover the whole ocean surface from latitudes of around 70°N to 70°S. The functioning generations of Inmarsat satellites comprise Inmarsat A, B, C, and E.

Inmarsat-A is an analog mobile satellite communication system coming to the end of its service life (Figure 2.12).

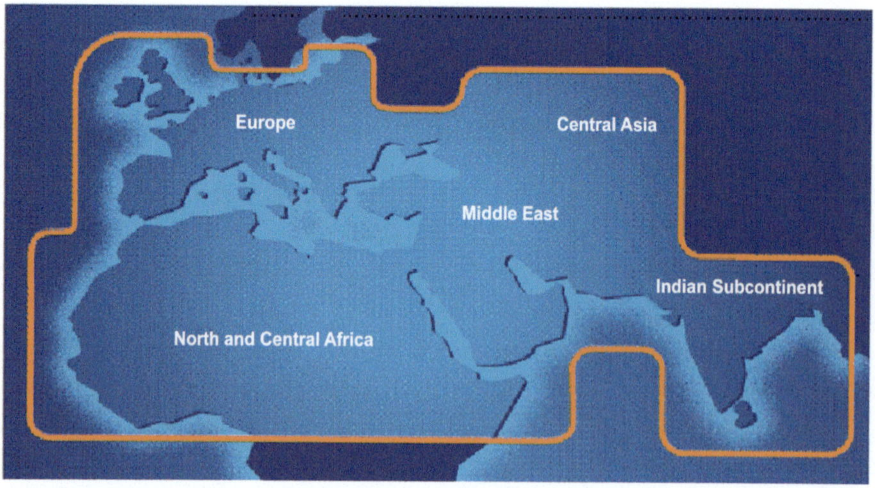

Figure 2.12. *The Thuraya coverage area.*

Source: https://slideplayer.com/slide/14028162/.

It accommodates 9.6 kbit/s data communications and voice in its normal configuration. An HSD (high-speed data) option presented by certain MES (mobile earth station) manufacturers allows data communications up to 64 kbit/s. Inmarsat-B compromises the digital replacement for the Inmarsat-A. It assists data, fax, voice, and telex. The standard data rate is 9.6 kbit/s; however, the supported rate for the HSD option is 64 kbit/s.

Inmarsat-C is a data store and forward messaging system and does not support voice. The data rate is 600 bits/s, and the extreme message length is 32 kB. The system could connect to data networks and telex. It delivers a facility for EGC (enhanced group calls), which allows groups of vessels in a geographic area to be instantaneously addressed.

Inmarsat-E comprises a built-in GPS receiver and geostationary satellites utilized mostly for rescue activities. (Gineste *et al.,* 2017).

REFERENCES

1. Ahedo, E., & Sanmartin, J. R., (2002). Analysis of bare-tether systems for deorbiting low-earth-orbit satellites. *Journal of Spacecraft and Rockets, 39*(2), 198–205.
2. Alagoz, F., & Gur, G., (2011). Energy efficiency and satellite networking: A holistic overview. *Proceedings of the IEEE, 99*(11), 1954–1979.
3. Alfriend, K. T., Lee, D. J., & Creamer, N. G., (2006). Optimal servicing of geosynchronous satellites. *Journal of Guidance, Control, and Dynamics, 29*(1), 203–206.
4. Araguz, C., Bou-Balust, E., & Alarcón, E., (2018). Applying autonomy to distributed satellite systems: Trends, challenges, and future prospects. *Systems Engineering, 21*(5), 401–416.
5. Baoyin, H., & McInnes, C. R., (2006). Trajectories to and from the Lagrange points and the primary body surfaces. *Journal of Guidance, Control, and Dynamics, 29*(4), 998–1003.
6. Bem, D. J., Wieckowski, T. W., & Zielinski, R. J., (2000). Broadband satellite systems. *IEEE Communications Surveys and Tutorials, 3*(1), 2–15.
7. Blumenthal, S. H., (2013). Medium earth orbit Ka-band satellite communications system. In: *MILCOM 2013–2013 IEEE Military Communications Conference* (Vol. 1, pp. 273–277). IEEE.
8. Broucke, R., (1979). Traveling between the Lagrange points and the moon. *Journal of Guidance and Control, 2*(4), 257–263.
9. Byers, D. C., & Dankanich, J. W., (2008). Geosynchronous-earth-orbit communication satellite deliveries with integrated electric propulsion. *Journal of Propulsion and Power, 24*(6), 1369–1375.
10. Chang, H. S., Kim, B. W., Lee, C. G., Min, S. L., Choi, Y., Yang, H. S., & Kim, C. S., (1998). FSA-based link assignment and routing in low-earth orbit satellite networks. *IEEE Transactions on Vehicular Technology, 47*(3), 1037–1048.
11. Cognion, R. L., (2014). Rotation rates of inactive satellites near geosynchronous earth orbit. In: *2014 AMOS Conference* (Vol. 1, pp. 9–12). Maui, HI.
12. Colino, R. R., (1977). International cooperation between communications satellite systems: An overview of current practices and future prospects. *J. Space L., 5*, 65.

13. Colombo, C., (2016). Planetary orbital dynamics (PlanODyn) suite for long-term propagation in perturbed environment. In: *6th International Conference on Astrodynamics Tools and Techniques (ICATT)* (Vol. 1, pp. 14–17).
14. Corazza, G. E., & Vatalaro, F., (1994). A statistical model for land mobile satellite channels and its application to nongeostationary orbit systems. *IEEE Transactions on Vehicular Technology, 43*(3), 738–742.
15. Corcoran, J. M., Knight, J. F., & Gallant, A. L., (2013). Influence of multi-source and multi-temporal remotely sensed and ancillary data on the accuracy of random forest classification of wetlands in Northern Minnesota. *Remote Sensing, 5*(7), 3212–3238.
16. De Weck, O. L., De Neufville, R., & Chaize, M., (2004). Staged deployment of communications satellite constellations in low earth orbit. *Journal of Aerospace Computing, Information, and Communication, 1*(3), 119–136.
17. Del, P. I., Cameron, B. G., & Crawley, E. F., (2019). A technical comparison of three low earth orbit satellite constellation systems to provide global broadband. *Acta Astronautica, 159*, 123–135.
18. Efimov, S., Pritykin, D., & Sidorenko, V., (2018). Long-term attitude dynamics of space debris in sun-synchronous orbits: Cassini cycles and chaotic stabilization. *Celestial Mechanics and Dynamical Astronomy, 130*(10), 1–25.
19. Ferguson, D. C., Hilmer, R. V., & Davis, V. A., (2015). Best geosynchronous earth orbit daytime spacecraft charging index. *Journal of Spacecraft and Rockets, 52*(2), 526–543.
20. Friesen, L. J., Jackson, IV. A. A., Zook, H. A., & Kessler, D. J., (1992). Analysis of orbital perturbations acting on objects in orbits near geosynchronous earth orbit. *Journal of Geophysical Research: Planets, 97*(E3), 3845–3863.
21. Gineste, M., Deleu, T., Cohen, M., Chuberre, N., Saravanan, V., Frascolla, V., & Dutkiewicz, E., (2017). Narrowband IoT service provision to 5G user equipment via a satellite component. In: *2017 IEEE Globecom Workshops (GC Wkshps)* (Vol. 1, pp. 1–4). IEEE.
22. Goldhirsh, J., (1992). *Propagation Effects for Land Mobile Satellite Systems: Overview of Experimental and Modeling Results, 1*, 1–33.
23. Gupta, O. P., (2011). Global augmentation of ADS-B using Iridium NEXT hosted payloads. In: *2011 Integrated Communications,*

Navigation, and Surveillance Conference Proceedings (pp. 1–15). IEEE.

24. Hikami, S., Larkin, A. I., & Nagaoka, Y., (1980). Spin-orbit interaction and magnetoresistance in the two-dimensional random system. *Progress of Theoretical Physics, 63*(2), 707–710.
25. Hu, Y., & Li, V. O., (2001). Satellite-based internet: A tutorial. *IEEE Communications Magazine, 39*(3), 154–162.
26. Huang, L., Qiu, X., Hu, D., & Ding, C., (2010). Focusing of medium-earth-orbit SAR with advanced nonlinear chirp scaling algorithm. *IEEE Transactions on Geoscience and Remote Sensing, 49*(1), 500–508.
27. Hudson, J. S., & Kolosa, D., (2020). Versatile on-orbit servicing mission design in geosynchronous earth orbit. *Journal of Spacecraft and Rockets, 57*(4), 844–850.
28. Jäggi, A., Hugentobler, U., Bock, H., & Beutler, G., (2007). Precise orbit determination for GRACE using undifferenced or doubly differenced GPS data. *Advances in Space Research, 39*(10), 1612–1619.
29. Kahr, E., OKeefe, K., & Montenbruck, O., (2014). Hardware-in-the-loop simulation of GNSS signal tracking in highly elliptical orbits using the GSNRx™ software receiver. In: *Proceedings of the 27th International Technical Meeting of the Satellite Division of The Institute of Navigation (ION GNSS+ 2014)* (Vol. 1, pp. 1448–1458).
30. Key, T., Warner, T. A., McGraw, J. B., & Fajvan, M. A., (2001). A comparison of multispectral and multitemporal information in high spatial resolution imagery for classification of individual tree species in a temperate hardwood forest. *Remote Sensing of Environment, 75*(1), 100–112.
31. Kirillov, A., (1999). Merits and demerits of the orbit method. *Bulletin of the American Mathematical Society, 36*(4), 433–488.
32. Kumar, K. R., Kang, N., & Yin, Y., (2018). Classification of key aerosol types and their frequency distributions based on satellite remote sensing data at an industrially polluted city in the Yangtze River Delta, China. *International Journal of Climatology, 38*(1), 320–336.
33. Lim, J., Klein, R., & Thatcher, J., (2005). Good technology, bad management: A case study of the satellite phone industry. *Journal of Information Technology Management, 16*(2), 48–55.
34. Logue, T. J., & Pelton, J., (2019). Overview of commercial small satellite systems in the "new space" age. *Handbook of Small Satellites:*

Technology, Design, Manufacture, Applications, Economics and Regulation, 1, 1–18.

35. Lücking, C., Colombo, C., & McInnes, C. R., (2012). A passive satellite deorbiting strategy for medium earth orbit using solar radiation pressure and the J2 effect. *Acta Astronautica, 77*, 197–206.
36. Luo, H., Mao, Y. D., Yu, Y., & Tang, Z. H., (2019). Focus GEO observations of space debris at geosynchronous earth orbit. *Advances in Space Research, 64*(2), 465–474.
37. Luu, B. B., & Sadler, C. M., (2001). Data transmission over Inmarsat in TCP/IP environment. In: *Digital Wireless Communication III* (Vol. 4395, pp. 233–239). International society for optics and photonics.
38. Lyras, N. K., Efrem, C. N., Kourogiorgas, C. I., & Panagopoulos, A. D., (2019). Medium earth orbit optical satellite communication networks: Ground terminals selection optimization based on the cloud-free line-of-sight statistics. *International Journal of Satellite Communications and Networking, 37*(4), 370–384.
39. Messenger, S. R., Wong, F., Hoang, B., Cress, C. D., Walters, R. J., Kluever, C. A., & Jones, G., (2014). Low-thrust geostationary transfer orbit (LT2GEO) radiation environment and associated solar array degradation modeling and ground testing. *IEEE Transactions on Nuclear Science, 61*(6), 3348–3355.
40. Morand, V., Le Fèvre, C., Lamy, A., Fraysse, H., & Deleflie, F., (2012). Dynamical properties of geostationary transfer orbits over long time scales: Consequences for mission analysis and lifetime estimation. In: *AIAA/AAS Astrodynamics Specialist Conference* (Vol. 1, pp. 4968).
41. Nicholson, J., & Gerstein, B. T., (2000). The department of defense's next-generation narrowband satellite communications system, the mobile user objective system (MUOS). In: *MILCOM 2000 Proceedings. 21st Century Military Communications. Architectures and Technologies for Information Superiority (Cat. No. 00CH37155)* (Vol. 2, pp. 805–809). IEEE.
42. Oh, H. U., Lee, K. J., & Jo, M. S., (2013). A passive launch and on-orbit vibration isolation system for the spaceborne cryocooler. *Aerospace Science and Technology, 28*(1), 324–331.
43. Oltrogge, D. L., Alfano, S., Law, C., Cacioni, A., & Kelso, T. S., (2018). A comprehensive assessment of collision likelihood in geosynchronous earth orbit. *Acta Astronautica, 147*, 316–345.

44. Paek, S. W., Kim, S., Kronig, L., & De Weck, O., (2020). Sun-synchronous repeat ground tracks and other useful orbits for future space missions. *The Aeronautical Journal, 124*(1276), 917–939.
45. Panagiotarakis, N., Maglogiannis, I., & Kormentzas, G., (2002). An overview of major satellite systems. In: *2nd WSEAS International Conference on Multimedia, Internet, and Video Technologies, 2*, 1–33.
46. Pratt, S. R., Raines, R. A., Fossa, C. E., & Temple, M. A., (1999). An operational and performance overview of the Iridium low earth orbit satellite system. *IEEE Communications Surveys, 2*(2), 2–10.
47. Richardson, R. W., (1972). Principal orbit types for algebraic transformation spaces in characteristic zero. *Inventiones Mathematicae, 16*(1, 2), 6–14.
48. Rossi, A., (2008). Resonant dynamics of medium earth orbits: Space debris issues. *Celestial Mechanics and Dynamical Astronomy, 100*(4), 267–286.
49. Ryden, K. A., Hands, A. D., Underwood, C. I., & Rodgers, D. J., (2015). Internal charging measurements in medium earth orbit using the SURF sensor: 2005–2014. *IEEE Transactions on Plasma Science, 43*(9), 3014–3020.
50. Sabetghadam, S., Alizadeh, O., Khoshsima, M., & Pierleoni, A., (2021). Aerosol properties, trends, and classification of key types over the middle east from satellite-derived atmospheric optical data. *Atmospheric Environment, 246*, 118100.
51. Sarigul-Klijn, N., Sarigul-Klijn, M., & Noel, C., (2005). Air-launching earth to orbit: Effects of launch conditions and vehicle aerodynamics. *Journal of Spacecraft and Rockets, 42*(3), 569–575.
52. Schöne, T., Pandoe, W., Mudita, I., Roemer, S., Illigner, J., Zech, C., & Galas, R., (2011). GPS water level measurements for Indonesia's Tsunami early warning system. *Natural Hazards and Earth System Sciences, 11*(3), 741–749.
53. Skoulidou, D. K., Rosengren, A. J., Tsiganis, K., & Voyatzis, G., (2019). Medium earth orbit dynamical survey and its use in passive debris removal. *Advances in Space Research, 63*(11), 3646–3674.
54. Sohier, H., Farges, J. L., & Piet-Lahanier, H., (2014). Improvement of the representativity of the Morris method for air-launch-to-orbit separation. *IFAC Proceedings Volumes, 47*(3), 7954–7959.
55. Sohier, H., Piet-Lahanier, H., & Farges, J. L., (2015). Analysis and

optimization of an air-launch-to-orbit separation. *Acta Astronautica, 108*, 18–29.

56. Tjelta, T., Nordbotten, A., Annoni, M., Scarrone, E., Bizzarri, S., Tokarchuk, L., & Dinis, M., (2001). Future broadband radio access systems for integrated services with flexible resource management. *IEEE Communications Magazine, 39*(8), 56–63.
57. Tomiyasu, K., (1978). Synthetic aperture radar in geosynchronous orbit. In: *Synthetic Aperture Radar Technology Conference* (Vol. 1, pp. 1–33).
58. Trishchenko, A. P., & Garand, L., (2011). Spatial and temporal sampling of polar regions from two-satellite system on Molniya orbit. *Journal of Atmospheric and Oceanic Technology, 28*(8), 977–992.
59. Trishchenko, A. P., & Garand, L., (2012). Observing polar regions from space: Advantages of a satellite system on a highly elliptical orbit versus a constellation of low earth polar orbiters. *Canadian Journal of Remote Sensing, 38*(1), 12–24.
60. Trishchenko, A. P., Garand, L., & Trichtchenko, L. D., (2019). Observing polar regions from space: Comparison between highly elliptical orbit and medium earth orbit constellations. *Journal of Atmospheric and Oceanic Technology, 36*(8), 1605–1621.
61. Tronc, J., Angeletti, P., Song, N., Haardt, M., Arendt, J., & Gallinaro, G., (2014). Overview and comparison of on-ground and on-board beamforming techniques in mobile satellite service applications. *International Journal of Satellite Communications and Networking, 32*(4), 291–308.
62. Verstraete, A. W., Anderson, D., St. Louis, N. M., & Hudson, J., (2018). Geosynchronous earth orbit robotic servicer mission design. *Journal of Spacecraft and Rockets, 55*(6), 1444–1452.
63. Wang, Y., & Gurfil, P., (2016). Dynamical modeling and lifetime analysis of geostationary transfer orbits. *Acta Astronautica, 128*, 262–276.
64. Wang, Y., Luo, X., & Wu, X., (2020). Long-term evolution and lifetime analysis of geostationary transfer orbits with solar radiation pressure. *Acta Astronautica, 175*, 405–420.
65. Weeden, B., & Shortt, K., (2008). Development of an architecture of sun-synchronous orbital slots to minimize conjunctions. In: *Space Ops 2008 Conference* (Vol. 1, pp. 3547).

66. Wilkinson, G. G., (2005). Results and implications of a study of fifteen years of satellite image classification experiments. *IEEE Transactions on Geoscience and Remote Sensing, 43*(3), 433–440.
67. Xu, W., Liang, B., Li, B., & Xu, Y., (2011). A universal on-orbit servicing system used in the geostationary orbit. *Advances in Space Research, 48*(1), 95–119.
68. Yan, L., Roy, D. P., Li, Z., Zhang, H. K., & Huang, H., (2018). Sentinel-2A multi-temporal misregistration characterization and an orbit-based sub-pixel registration methodology. *Remote Sensing of Environment, 215*, 495–506.
69. Yang, L., Ye, Z. Y., & Wu, J., (2016). The influence of the elastic vibration of the carrier to the aerodynamics of the external store in air-launch-to-orbit process. *Acta Astronautica, 128*, 440–454.
70. Yizhou, H., Gaofeng, C., Pengxu, L., Ruijun, C., & Weidong, W., (2015). Random access preamble design based on time pre-compensation for LTE-satellite system. *The Journal of China Universities of Posts and Telecommunications, 22*(3), 64–73.

Chapter

3 INTRODUCTION TO LOW EARTH ORBITAL (LEO) SATELLITES

CONTENT

3.1. Introduction 72

3.2. Leos Economic Viability 74

3.3. Leos Research Issues 76

3.4. Summary and Discussion 88

References 90

3.1. INTRODUCTION

Revolutionary changes are being observed in telecommunication systems. These revolutions have changed how service and industrial organizations operate, are transforming society, and are profoundly affecting the daily life of everyone. Communication systems based on low earth orbit satellite (LEOS), are an exciting and fascinating endeavor in reorganizing the services that global communication network provides. Out of many systems under development, a few are Orbcomm, Iridium, Teledesic, and Globalstar. All the systems target the masses and aim to provide them global communication services (Fossa et al., 1998; Del Portillo et al., 2019). In managing and developing mobile communication systems based on a large-scale commercial satellite, very limited experience exists; this makes it an extremely risky business. LEOS-based communication systems are presented in this chapter. Here, we will analyze their economic viability and will also discuss some of the potential research areas that are involved in their configuration, development, operation, management, and maintenance (Ware et al., 1996; McDowell, 2020).

LEOS providing global communication services is one of the exciting and new development. LEOS systems are based on the concept of having multiple satellites that orbit in low orbits. They have sophisticated equipment for transmitting, processing through antennas, communicating to and from hand-held user terminals present on the ground. As implied by low earth satellite orbits, the satellites move continuously relative to the ground, disappearing, and again appearing from the user's sight, compelling recurring hand-offs of users among beams of the antenna within a satellite and also from present satellite to upcoming serving satellite (Pratt et al., 1999).

It is expected that LEOS will be providing wireless mobile communication services around the world. Out of many advantages they have, one is their transcending ability; they transcend the artificial boundaries imposed by regional, state, and local governing bodies. LEOS capably provide instant communication services in regions where telecommunication infrastructures are missing or are underdeveloped, i.e., South America, Africa, Asia, and Eastern Europe. These systems will be supporting wireless communication in areas that are not covered by geostationary or cellular phone systems, i.e., earth poles, wilderness areas, deserts, or oceans. They will be supporting a huge range of services, such as data communications, phone, paging, messaging, data communications, video services, broadcasting, positioning, monitoring, narrowband, and wideband broadcasting and communication services (Chang *et al.,* 1998).

In the deployment, assembly, and development of LEOS, significant investments are made. Ongoing operational costs that range to several billion dollars/year will be incurred by LEOS. This cost is required to ensure that the operation continues, to replace dead satellites, and for management and marketing costs. It is expected that LEOS will complement and support the global communication system's other components as well, such as the wireless system that consists of data communication and cellular phone services and geostationary satellites, and wire-based (fiber and copper) system.

LEOS system's basic components include hand-held communication devices, satellites orbiting in low orbits (mostly the altitudes are between 700 and 5000 kilometers), and gateways to and from the ground-based communication systems. Satellites in low orbits implicit that the satellites, relative to the Earth, are not stationary; they keep moving while staying in their orbits with the rotation time in between 100 and 120 minutes/rotation. This time depends on their altitude and trajectory. For ensuring that every single point receives continuous coverage as well as communication, at least one satellite has to be above the minimum threshold of angular elevation level (at any point of interest) and within the line of sight (LOS). For a particular system, the selected satellite configuration has to provide uninterrupted service and coverage even under external interference or components and satellite failure conditions (Leo and Brown, 2000).

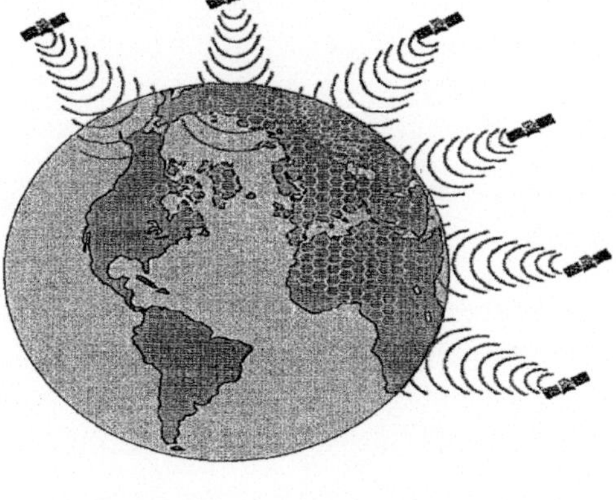

Figure 3.1. *A global LEOS communication system.*

Source: https://ishare.iask.sina.com.cn/f/61930973.html.

Numerous systems based on LEOS are under development. The systems have different characteristics (numerous satellites, communication technologies, satellite weight, constellations, antenna types, trajectories, types of offered communication services). Different approaches to using these new technologies are represented by them. Significant investments are required by each of these systems. Moreover, being an untested and new technology, system designers have to address many operational and technical challenges to make these operations successful and economically viable (Figure 3.1) (Leo, 1998).

This chapter aims to introduce the major problems and present their initial results to grab the research opportunities in the potential fields. The LEOS consortium has undertaken all extensive development and research efforts. Unfortunately, because of their high financial risks and the novelty involved, similar companies protect such models as commercial secrets. Their publication is not encouraged. Recently, details of some models have been published in several papers. Wherever appropriate, we refer to public domain papers containing both the analysis and modeling aspects of LEOS. Early discussions regarding LEOS analysis issues can be seen in Gavish (1995a, b).

3.2. LEOS ECONOMIC VIABILITY

In global communication systems, LEOS presents innovative and new development. With this unproved and new technology, investors face many risks. Mostly the question that arises is: Whether the designed systems will ever be successful or not in the marketplace? The following issues should be addressed by any technologically risky and new system (Leo *et al.,* 1998; Díaz-de-Baldasano *et al.*, 2014):

1. Whether the system is technologically feasible or not? Irrespective of the revenues collected and costs by the system, can we find a feasible design?
2. What are the managerial and political considerations, and what sort of environment supports them? The frequency allocation by the regulatory agencies for a particular service or system can be taken as an example of discussed geopolitical consideration.

For minimizing the cannibalization of existing services, incentives such as assurances are required. These existing services can be cellular phone services; they can be orders of materials or other services that suppliers offer, especially in countries that support the project. Questions that arise are:

Who can become a shareholder or can invest in the system? The satellites will be launched by whom? Who will be responsible for manufacturing the different satellite components? Who will be launching the satellites? At what percentage will the revenues be shared?

3. The system's economic viability: After such a huge investment, will the return be high enough that it will be able to provide the system's long-term economic viability? Will the price of services (that user will be paying) be low enough that it attracts a larger user community so that the system operation and development costs can be recovered?

For analyzing the system's economic viability during its lifetime, the system's basic cost components are identified. For better comparison, the cost of the main system under steady-state conditions is analyzed. For example, the Iridium configuration is used to calculate the system's annual lifetime costs.

i. **Satellite Replacement and Launch Costs:** For the Iridium system, as per the plan, there should be 66 operational satellites in orbit. In-orbit satellites have a limited lifetime, and due to this, it is expected that satellites had to be launched into orbit in a steady state to replace dying or dead satellites. Considering the expected lifetime of a satellite to be of 5 years, we can calculate the average number of satellites needed to be replaced yearly, which is 66/5=13.2. Satellites and rockets are not perfect; hence there exists a possibility that they might fail after or during the launch process. The success probability for commercial rockets is between 0.8 and 0.96 (Gavish and Kalvenes, 1995, 1997b). This probability depends on the launch method, cost, payload, and rocket type. Considering the launch failures after and during launch may lead to the need of launching at least 15 satellites per year (Andriulli et al., 2004).

ii. **Launching Cost:** It is estimated while keeping 15 satellite launches/year, for 10 million dollars for each launch. So, in total, $150 million is estimated for each year.

iii. **Satellite Replacement Cost:** It is also estimated while keeping 15 satellites in consideration, so almost 20 million dollars for one satellite. So, in total, $300 million is estimated for each year.

iv. **Operation and Gateways Costs:** These are estimated at $100 million each year.

v. **Billing, Marketing, Management, and Accounting Costs:** These are estimated at $100 million each year.

4. The system's largest cost component is constituted by *financing charges*. It includes the recovery of development and research costs, financing the inventory of the ground satellites and of the ones in orbit, gateways, rockets, and spare parts required to keep the system running and up.

Gathering the different cost components, the total cost sums up to $950 million per year. Assuming that this system will be able to achieve 1,000,000 subscribers, still, each one of them will have to pay $1000 each year to keep it going for the long run. The Iridium system has declared to achieve a target of 10,000,000 subscribers, then in this ideal case, each one of them will be paying almost $10 each month (for recovering the annual costs). If the subscribers are charged 10 dollars each month, this will raise the number of users, making the system attractive for everyone. This might even exceed the demand to the point that it exceeds the system's capacity. Based on these considerations, the Iridium system seems to solve political and technical questions. If this system reaches its stated goal (the number of subscribers), it will turn out to be a huge success.

3.3. LEOS RESEARCH ISSUES

There is a huge profit potential in services based on LEOS, and this has attracted investments and attention by large international concerns. For having a proper understanding of the promised benefits, many operational and technical questions have to be answered. These systems provide fertile and new grounds for research-related activities.

3.3.1. Constellation Configuration

There is an impact of the number of orbits and satellites present in them, orbit altitude, and orbits type on overall operating costs and system configuration. When designing a constellation for these systems, the main objective is to ensure that all the satellites stay within the line of sight (LOS). These satellites have to be within the line of the interested service points on Earth. Constellations might vary from rosette constellations for polar-based trajectories to different combinations of various configurations. A few of the commonly used configuration's classes include (Hongzheng and Chao, 2011):

1. Polar orbits-such constellations where the orbital planes (relative to the poles) have a slight inclination or pass over the poles. Each satellite is in a position that is highly predictable, which simplifies the communication control structures required for the system. For the regions close to the poles, a high coverage degree is provided by polar orbits. This high degree of coverage is regarded as a major disadvantage. However, over the globe, considering the actual termination locations of communicating parties and distribution of origination, we can conclude that polar orbits are advantageous when we incorporate power management issues as the design factor. In regions where high-power consumption and traffic are expected, multiple coverages are provided by polar orbits.
2. Constellations, which have highly inclined orbits relative to the equator are termed Rosette constellations. They provide a high coverage level for all parts of the globe except for polar areas. In polar areas, they provide a lower coverage level. In a rosette constellation, it is difficult to maintain interstate links. Any of the LEOS (announced so far) has not proposed the use of a rosette constellation.
3. Another constellation is Equatorial constellations. It provides excellent coverage at the equator but offers no coverage in areas that are away from the equator.
4. A minimum number of satellites are required by Polyhedral orbits to offer continuous global coverage. Polyhedral orbits achieve this at the expense of higher satellite altitudes and complicated orbits. They have complex orbital structures, hence maintaining up and down communication links and supporting space-based routing is difficult. It makes polyhedral orbits unattractive for the systems based on LEOS.

Depending upon many design factors, the constellations from above are selected. Some designs use a combination of the above configurations. For a LEOS system, three possible constellations are shown in Figure 3.2: a rosette constellation, a polar constellation, and the one consisting of an equatorial and rosette constellation, named as a mixed constellation. The type of constellation affects both, the number of satellites required to provide full earth coverage and the launch costs of satellites that are required for the initial configuration, as well as for the maintenance of the system (Fallon and Oestreich, 2015).

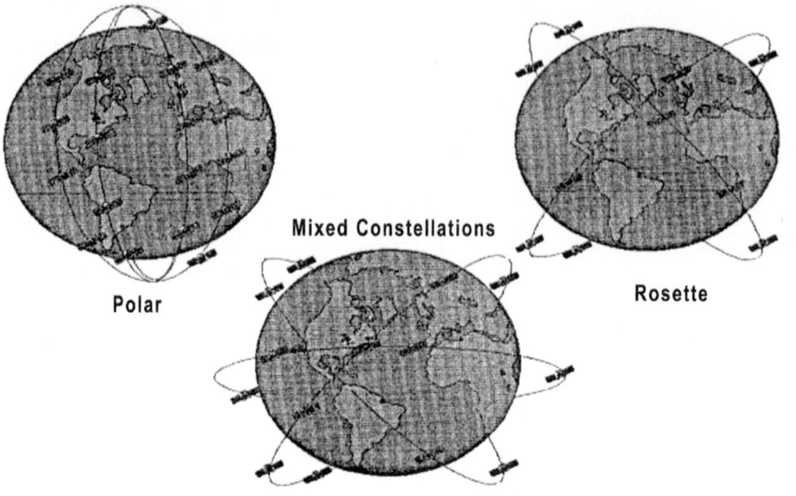

Figure 3.2. *Examples of satellite constellations.*

Source: https://www.sciencedirect.com/science/article/abs/pii/S0377221796003906

The early studies regarding configurations guaranteeing a coverage level encompass the papers by Beste (1978), Ballard (1980), and Perrotta (1991). Beste used classical optimization methods for finding the minimum number of satellites required for both multiple and single coverage in polar orbits. Ballard studied rosette constellations. Studied rosette constellations compared to elliptical and circular orbits. Other papers related to trajectory design include Kaniyil et al. (1992); and Adams and Rider (1987). The used nonlinear optimization for designing polyhedral orbits that may have a minimum number of satellites to provide complete coverage of the Earth. Furthermore, other investigators include Rider (1985, 1986); Maral et al. (1991); Markowic and Hope (1992); and Sheriff and Gardiner (1993).

The cost of offering a stated level of coverage is not provided by the above-discussed studies. Higher coverage will result in higher system costs We can find such an example in the Iridium system, where the in-orbit satellites were reduced to 66 (instead of 77) satellites at the cost of few minutes every 24 hours by not covering the areas that were near to the equator. By doing this, around 15% of system cost was reduced. It is purely an example of a worthwhile trade-off.

In selecting any configuration, the following decisions are included (Shi et al., 2018):

1. **The Number of In-Orbit Satellites:** The cost (finance) of the system is increased by a higher number of satellites. It also increases the inventory and in-orbit system power reserve.
2. **The Number of Orbital Planes:** If the number of orbits is higher, the distance among satellites adjacent orbits decreases, resulting in reducing the energy requirements and signal propagation time of inter-satellite links.
3. **The Inclination of Orbital Planes:** If the orbits are more inclined, the likelihood of satellite collisions are observed to be low when they pass near or over the poles.
4. **The Orbital Plane's Angular Spacing:** It determines the cross-seam distance among orbits.
5. **The Number of Satellites Present in Each Orbital Plane:** If a high number of satellites are present, the in-orbit propagation times are decreased between adjacent satellites.
6. **Satellite's Relative Spacing:** This is within an orbital plane.
7. **Satellite's Angular Inclination:** This is among adjacent orbital planes.
8. **Level of Coverage:** Several planned systems, i.e., Teledesic, needs multiple coverages of the terminal. It is required so that the satellites can effectively operate their communication system.
9. **Storage Potential and Power Collection:** A nonlinear model was developed by Gavish and Kalvenes (1996) that links power generation, satellite altitude, and storage to the total weight that is allocated for power generation and storage on a satellite concerning the overall system capacity. By capacity, we refer to the number of calls that the system supports. The phone call duration is also provided by the same model. Their calculations demonstrate that in configuring LEOS systems, the power storage capacity for each unit of weight is an essential factor (Li et al., 2010).
10. **Satellite Altitude:** A much high degree of frequency reusability is implied by a lower altitude. However, the atmospheric drag is increased on the satellite by a lower altitude. On the other hand, the expected useful lifetime of a satellite is decreased. The satellite launch cost is also reduced (but as the orbit lifetime of the satellite is reduced,

the per-day launch cost may increase). The number of satellites required for covering the Earth is decreased by a higher altitude. Contrary to it, higher altitude increases the power requirements for the equipment transmitting it, weight of the satellite, the satellite launch costs, and it may also decrease the lifetime of equipment present in the satellites (this is because of ionosphere provides lower protection level and the Van Allen effect).

11. **Elliptic vs. Circular Trajectories:** From points on Earth, small changes in altitude are offered by circular trajectories. This makes the signal acquisition and positioning simpler. Intersatellite communications within orbit are also simplified by circular orbits (Budianto and Olds, 2004).

3.3.2. Physical Satellite Configuration

It includes the types and numbers of antennas utilized in direct user communication (multibeam/spot beam/ single); movable antennas vs. fixed ones for up and downlink communications; the types and numbers of antennas used in gateway communication; the type (optical or electromagnetic) and the number of inter-satellite communication links that will provide support (the inter-satellite links under development in the systems are in between 0 and 8; Figure 3.3 shows an example of a system having eight inter-satellite links named as the Teledesic design); energy storage devices along with their types; energy collection surface areas; satellite propulsion/ maneuvering subsystems; energy collection control mechanisms; and switching, receiving multiplexing, and transmission technologies (Girard et al., 2015; Radhakrishnan et al., 2016).

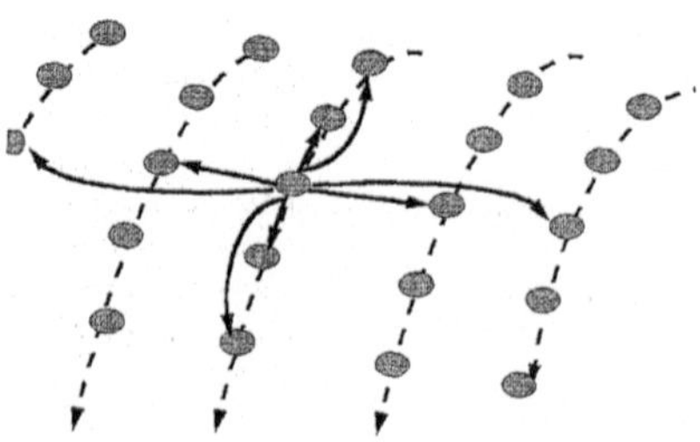

Figure 3.3. *Example of a satellite along with its immediate eight inter-satellite links.*

Source: https://ideas.repec.org/a/eee/ejores/v99y1997i1p166-179.html.

The design decisions are intertwined. Mostly, the technical specifications of many components directly affect the design of other components. Considering the limited lifetime of some components (such as solar panels), it is of no use to design other satellite components for a time that is far greater than the expected lifetime of the components having a shorter lifetime (Smith et al., 1992; Powell et al., 2006). Corresponding to it, redundancies have to be included so that a failure of one component does not result in a complete satellite failure.

3.3.3. Intersatellite Links

The forthcoming issue is the operation and configuration of inter-satellite links. Should the system support space-based routing? On earth-based services, the dependence of the system is reduced by space-based routing provided by telcos. For space-based routing, various technologies are possible through inter-satellite links, which add up to the power and weight requirements of the systems (Werner *et a*l., 1997; Wang *et al*., 2019). They differ in the maximal angular change rate among the satellite's relative positions and their distance, which they can support. Satellite interconnection's different possible patterns are shown in Figure 3.4. These patterns head to various end-to-end means and sometimes may lead to worst-case delays affecting the overall performance. In LEOS systems, communication links between adjacent satellites (same-orbit). It uses polar circular orbits. As within the orbit, the relative positions of satellites are not changing, and it is easy to support communication links having polar circular orbits. In adjacent orbits, inter-satellite links between satellites are comparatively difficult to support. Within and between elliptic orbits, inter-satellite links are very extremely difficult to support; hence, most of the systems planning to use elliptic orbits depend mostly on ground-based routing (Bertossi *et a*l., 1987).

In space-based routing, a prominent issue to be addressed are cross-seam links. Between the orbits, seams are formed when satellites move in opposite directions in two adjacent orbits. Among polar-based systems, this happens twice. Communication between satellites moving over the seams (in opposite directions) is comparatively difficult to support. We can handle cross-seam communications by routing their messages to the other side (over the pole), two satellites moving in the same direction. Long propagation delays and various hops in the routes are implied by such over-the-pole routing. Limited research regarding the inter-satellite link technology's impact, crosslink operational policies, and crosslink configuration patterns on overall system performance are published. Models of inter-satellite links were developed by Gavish and Kalvenes (1997a). They used this model for analyzing the effect of various crosslink configuration patterns. Gavish and Kalvenes used shortest path routing models and also calculated the worst case and overall end-to-end delay. The end-to-end delay and worst-case are illustrated in Figure 3.5. In Figure 3.4, pattern B of crosslink configuration, about the end-to-end delay, is more preferred than other tested patterns (Wu *et al.*, 2014).

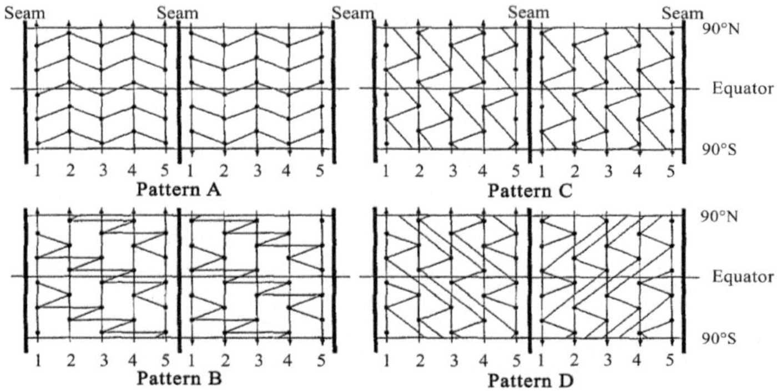

Figure 3.4. *Different satellite interconnection patterns.*

Source: https://www.sciencedirect.com/science/article/abs/pii/S0377221796003906.

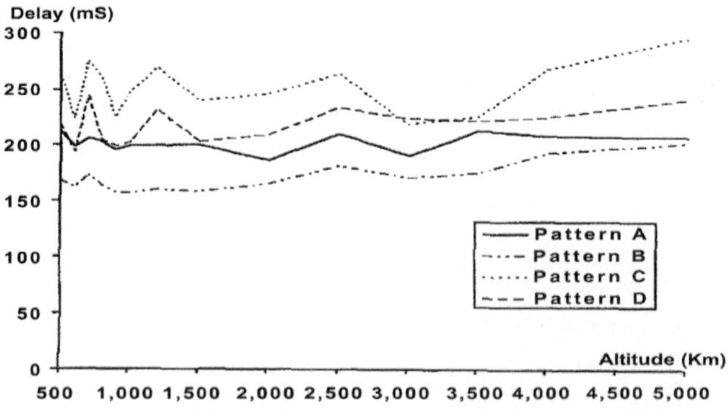

Figure 3.5. *Worst-case delay. It is a function of satellite altitude and crosslink patterns.*

Source: https://onlinelibrary.wiley.com/doi/abs/10.1002/sat.4600090403.

Surprisingly, all the systems based on polar orbit use an inferior pattern (pattern A). Using a combination of numerical simulation techniques and shortest path algorithms (see Lawler, 1976; Bertsekas, 1995a), Gavish and Kalvenes also investigated the effect of using expensive technologies to delay the crosslink's shutoff point to higher latitudes, and the cross-seam links effect that supports direct communications among satellites. The analysis gives insight into the economic futility (or worthiness) of investing in expensive and new technologies to extend inter-satellite communications (Newman *et al.,* 2012).

3.3.4. Communication Methods and Services

What transmission methodologies or their combinations should be preferred by the channels? What sort of communication services shall be provided/ supported by the system? All these services add to the system management complexity, software, and hardware, and the revenues and costs collected. For each of these services, appropriate software and equipment need to be installed on both the satellite and ground sides. What services given LEOS configuration should offer and the interaction between system configuration and service capabilities are interesting subjects of research for LEOS potential providers and designers (Mikkonen *et al.,* 2002; Hinami *et a*l., 2009).

3.3.5. Routing Methodologies-Space vs. Ground-Based Routing

Orbits are moving relative to the Earth, while satellites move within orbits. Inter satellite links, beams, satellites, and antennas may be switched off or again, depending on many physical and operational constraints. A combination of earth-based components and space-based components may compose routes. End-to-end communications with bounded variability and an acceptable delay, under many failures and operational conditions, should be ensured by the system. Considering all this, the fact of developing new routing methodologies cannot be denied (Bhalaji, 2019). These methodologies should be reliable and robust in a dynamic environment. In LEOS-based systems, routing has to provide sustenance for continuously changing topologies and stochastic demand.

In a LEOS system, each satellite covers a limited area at any point. Two distant communicating entities are covered by different satellites. Resultantly, it is required to route the message to a destination satellite (from an origin satellite). For interconnecting the destination and source points, LEOS architects have suggested two competing approaches. The first approach depends on the cable system (existing on the ground) for routing the message. In the routing approach based on Earth, assuming the communicating entities to be dependent on satellite communications, a source user interacts with a satellite, relaying its message to a gateway present on the earth station. At this point, the ground-based wire plant is used for transferring those messages to another gateway that is present near the destination point. This point beams the message to the other satellite and finally, this satellite sends it to the destination (Zaitchik *et al.*, 2010).

Numerous concerns have to be taken into consideration while designing, and these concerns include the stochastic user demand, the ground (gateway routing and operation) based charges dependent on ground-based operator and gateway, the possibility of component and satellite failures. Location models like the ones developed by Deng and Simchi-Levi (1991), Bitran *et al.* (1981), and Cattrysse and Van Wassenhove (1992) can be used for addressing a few of the issues regarding the gateway placement.

For interconnecting two communicators, the second approach uses communication links based on inter-satellite space, in order for transferring messages directly between satellites. Figure 3.6 depicts space-based vs. earth (or ground) based routing.

As compared to space-based routing, ground-based routing is less complex (technically). It is dependent on inter-satellite links. For satellites, the stabilization of inter-satellite links is relatively easy in the same circular orbit. This is because the satellites' relative positions do not change over time (Lienig et al., 2002). In different orbits, it is technically challenging to support inter-satellite links between satellites. Spaced-based routing has a major advantage; its system is self-contained, so it does not depend on services that are being provided by organizations such as independent and regional phone companies and PTTs). Political independence is increased in the case of using space-based routing. Taking the operational side into consideration, multi-objective routing models have to be developed. An example of such a model can be the one in Henig (1986, 1984). Different criteria have to be balanced, i.e., satellite's power consumption, end-to-end delay, revenues, and costs of the entities that are involved in the quality of service provided and the particular route. Methods such as multi-objective routing and multi-criteria optimization play an essential part in such routing problems. In LEO systems, a more complete analysis and exposition of routing issues are presented in Gavish (1997b).

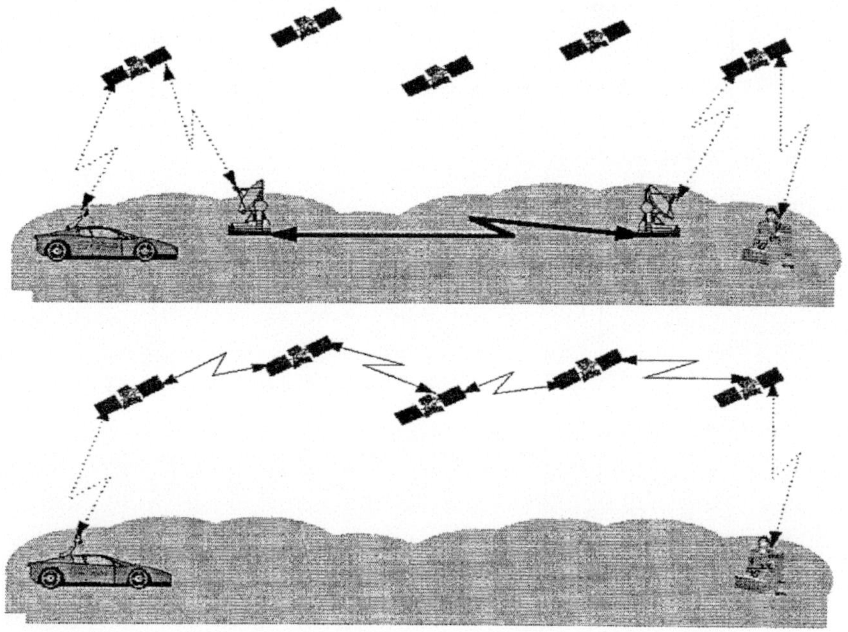

Figure 3.6. *Spaced-based vs. ground-based routing.*

3.3.6. Bandwidth Management and Allocation of Channel

Bandwidth serves as a limiting agent in the system's capacity. For maintaining the system economically viable, efficient frequency plays an essential part. The most acknowledgeable feature of such systems is high-level channel reusability. In highly dynamic environments, LEOS may cause challenging channel allocation problems. The very high altitude and speed of the satellites (compared to Earth) give rise to physical constraints. These constraints need to be addressed when allocating channels. Timeshift between satellites and within a satellite and the Doppler effects impose channel separations between satellites and between adjacent 'cells' (Del Re et al., 1995; Luo and Ansari, 2005). For LEOS systems, significant frequency shifts might be observed, and this is because of the planned frequency ranges and satellite velocities for communication systems. A satellite passes over the total area of the cell in a time-varying from a minimum of few seconds to a maximum of one minute. Different satellites may provide multiple coverages to cells, requiring a decision that which satellite will serve the cell at that point (Shah *et al.,* 2005; Al-Mistarihi *et al.*, 2012). For LEOS-based systems, the essentially important thing for its successful operation is efficient channel management. As a solution to these channel allocation problems, a vital role is played by models such as capacitated fixed charge networks (Wolsey, 1989) and dynamic graph coloring (taking place in real-time).

3.3.7. Power Management

Satellites' power consumption is a complex function of numerous factors, including stochastic/ fluctuating demand for satellite orbit and telecommunications, weather conditions, satellite household keeping operations, and demand for inter-satellite communications. The solar panels present on a satellite need to be always oriented toward the sun. This ensures the collection of optimal energy and prevents the panels from burning. During their orbits, satellites pass through the shadow of the Earth, and at that time, sunlight does not fall on their solar panels. There are limited energy collection areas on a solar panel of a satellite (Mostacciuolo et al., 2018). As there are weight limitations, the small size of the energy storage capacity of a bounded battery. The energy is consumed by the satellite transmission activities, and this may deplete its energy sources. If in case, the stored energy is depleted, the satellite is no longer useful, and generally, it cannot be reactivated. In the system operation, satellite activity management for conserving its energy is a crucial factor. It is possible to conserve energy by dividing the tasks of satellites and assigning them to other satellites, or by reducing active phone sessions that a satellite handles, or by shutting off inter-satellite connections or gateway connections, or multi-beam antennas. The operations of power management are handled by simulating many power consumption and storage scenarios, and through testing simple decision rules for managing the system. Further investigation needs to be performed to do this task most efficiently. A combination of stochastic optimization methods (Bertsekas, 1995b; Bellman and Dreyfus, 1962; Howard, 1960) and stochastic control models will be required to form effective power management procedures (Falke et al., 2004).

3.3.8. System Capacity

LEOS system investors and designers are concerned about this issue. System capacity determines the number of users that can appropriately use the system, while the quality of service is acceptable. The numbers of effective users determine the number of subscribers that can be efficiently accepted by the service providers. This affects the cost of charges for each subscription and for the use of the system (Alvarez and Walls, 2016; Chin *et al.,* 2018).

Very few researches regarding projected system capacity are published. The basic reason behind this might be that there are numerous factors responsible for determining the capacity of the system. Some of the impacting factors include; channel reusability level, control policies, regulatory bodies allocating frequency, transmission methods, demand patterns for communications while considering their shifts over a 24 hrs cycle, antenna technologies, and power consumption. On the overall system, the power consumption limit's effect was investigated by Gavish and Kalvenes (1997a).

We can reduce the dependency of system capacity on generating and storing power by increasing the number of satellites present in orbit. The planned Teledesic system demonstrates this reducing power dependency approach. It sets up 840 in-orbit satellites, which are responsible for generating enough energy that can drive the system efficiently at its best of theoretical capacity (Radhakrishnan *et al.*, 2016).

3.3.9. System Availability and Reliability

It is possible that with time, different parts of a satellite may undergo failure. As commonly practiced in ground-based systems, we cannot repair the hardware failures by simply sending a repair crew that will replace the failed component. Satellite and system designs should possess built-in redundancies, which should be capable enough to cope with in-orbit failures. As there are constraints such as the satellite weight and system cost, only a limited level of redundancy can be added to the system (Crisp *et al.*, 2014, 2015). The question that may arise is that where such redundancy should be built that subjects to volume constraints, weight, satellite lifetime distribution, power, and budget. Other questions regarding operational issues that may arise are: How the component failures should be handled? What should be the capacity of the system if it encounters different failure conditions? What should be the performance of the system under failing conditions? It should be kept in mind that most of the time, failure, it is meant that the quality of provided service is degraded (not that a system or satellite is shut-off). A combination of economic models and reliability models (Shogan, 1976; Li and Silvester, 1986; Ball, 1979) will be needed to address system availability and reliability issues.

3.3.10. Satellite Replacement and Launch Policies

In a LEOS system, the satellites have a limited lifetime. This lifetime comes from two major sources. The low altitude of the satellite imparts that gravity and drag will attract the satellite towards the Earth and will finally burn it in the atmosphere. The more appropriate source responsible for the short lifetime of the satellite is the eventual propellant depletion required for maneuvering the satellite. The satellites in LEOS have an expected lifetime ranging between 5 and 8 years (Cornara et al., 1999). Considering the satellite launch vehicle's limited capacity, the probability of launching failures, in-orbit shortage level of rocket, satellites, and expected revenue losses and satellite costs, the aim is to look for the optimal satellite replacement/ launch policies. For the static case, Gavish and Kalvenes (1997b) addressed this problem. They assumed that they would be already aware of the satellite shortages. Dynamic programming procedures were used by them for computing the optimal satellite launch policies. They also used dominance rules for reducing the exponential state space down to a size that can easily be managed. In Gavish and Kalvenes (1995), the static assumptions are less strict. A difference of 10 million dollars per anum was demonstrated while comparing different satellite replacement and launch policies. This was done through stochastic optimal control procedures. In Gavish and Kalvenes (1997b), an interesting question investigated was regarding dark satellites. The dark satellites may be kept parked in space, and when active satellites fail, the dark satellites can be moved from parking orbits and can be activated as replacement satellites (Jakob *et al.,* 2019).

3.4. SUMMARY AND DISCUSSION

There are many research issues, out of which we have introduced and discussed a few involved in communication systems that are based on LEOS. The future operators and designers of LEOS face many challenging questions. Most of the operational and design problems are difficult to solve and are NP-complete (Garey and Johnson, 1979). LEOS systems are a novice, so they provide (and will continue to do that) a fertile ground to numerous researchers interested in this emerging and potential field.

Systems based on the geostationary satellite are being used and are providing communication services for around three decades. The services they provide include security services, TV broadcasting, collection of sensing data, VSAT based data communication, monitoring, limited phone service, and paging. Communication systems based on MEOS (at 10,000 to 15,000 km altitude), medium earth orbit (MEO) satellite, have been put forward as a superior alternative to geostationary and LEOS satellites; they include systems like ICO and Odyssey. The role they play depends on the services and functions offered to the users. We have highlighted a few disadvantages and advantages of each system in Gavish (1997a), and have shown their operational characteristics affecting the choices made concerning the system.

Mobile communication services have been greatly facilitated by cellular communication networks (both satellite and terrestrial). For communication services, besides the regular variations, mobile systems also undergo stochastic changes, which are due to the mobility of customers. In configuring such systems, the added variability raises new challenges. The global reach of such systems (satellite-based) adds many administrative and political considerations to the economic and engineering aspects of these systems. For instance, for meeting the revenue targets promised to governmental agencies or PTTs, countries impose operational restrictions over which the satellites pass, in-demand time, and spatial changes for telecommunication services. Taking this type of factor into consideration for the day-to-day operations of the system, it becomes a complex task (Maier *et al*., 2018).

REFERENCES

1. Adams, W. S., & Rider, L., (1987). Circular polar constellations providing continuous single or multiple coverage above a specified latitude. *Journal of the Astronautical Sciences, 35*(2), 111–117.
2. Al-Mistarihi, M. F., Mowafi, M. Y., & Marei, M. S., (2012). An optimum scheme for WiMAX bandwidth management using dynamic channel allocation. In: *2012 Proceedings of the 35th International Convention MIPRO* (Vol. 1, pp. 538–541). IEEE.
3. Alvarez, J., & Walls, B., (2016). Constellations, clusters, and communication technology: Expanding small satellite access to space. In: *2016 IEEE Aerospace Conference* (Vol. 1, pp. 1–11). IEEE.
4. Andriulli, A., Solmi, L., Loperfido, S., Leo, P., Festa, V., Belmonte, A., & Perri, F., (2004). Prophylaxis of ERCP-related pancreatitis: A randomized, controlled trial of somatostatin and gabexate mesylate. *Clinical Gastroenterology and Hepatology, 2*(8), 713–718.
5. Ball, M., (1979). Computing network reliability. *Operations Research, 37*, 823–838.
6. Ballard, A. H., (1980). Rosette constellations of earth satellites. *IEEE Transactions on Aerospace and Electronic Systems, 16*(5), 656–673.
7. Bellman, R., & Dreyfus, S. E., (1962). *Applied Dynamic Programming* (Vol. 1, pp. 1–23). Princeton University Press, Princeton, NJ.
8. Bertossi, A. A., Bongiovanni, G., & Bonuccelli, M., (1987). Time slot assignment in SS/TDMA systems with inter-satellite links. *IEEE Transactions on Communications, 35*(6), 602–608.
9. Bertsekas, D. P., (1995a). *Dynamic Programming and Optimal Control* (Vol. I, pp. 1–24). Athena Scientific, Belmont, MA.
10. Bertsekas, D. P., (1995b). *Dynamic Programming and Optimal Control* (Vol. II, pp. 1–44). Athena Scientific, Belmont, MA.
11. Beste, D. C., (1978). Design of satellite constellations for optimal continuous coverage. *IEEE Transactions on Aerospace and Electronic Systems, 14*(3), 466–473.
12. Bhalaji, N., (2019). Performance evaluation of flying wireless network with Vanet routing protocol. *Journal of ISMAC, 1*(01), 56–71.
13. Bitran, G. R., Chandru, V., Sempolinski, D. E., & Shapiro, J. F., (1981). Inverse optimization: An application to the capacitated plant problem. *Management Science, 27*, 1120–1141.

14. Budianto, I. A., & Olds, J. R., (2004). Design and deployment of a satellite constellation using collaborative optimization. *Journal of Spacecraft and Rockets, 41*(6), 956–963.
15. Cattrysse, D. G., & Van, W. L., (1992). A survey of algorithms for the generalized assignment problem. *European Journal of Operational Research, 60*, 260–272.
16. Chang, H. S., Kim, B. W., Lee, C. G., Min, S. L., Choi, Y., Yang, H. S., & Kim, C. S., (1998). FSA-based link assignment and routing in low-earth orbit satellite networks. *IEEE Transactions on Vehicular Technology, 47*(3), 1037–1048.
17. Chin, K. B., Brandon, E. J., Bugga, R. V., Smart, M. C., Jones, S. C., Krause, F. C., & Bolotin, G. G., (2018). Energy storage technologies for small satellite applications. *Proceedings of the IEEE, 106*(3), 419–428.
18. Cornara, S., Beech, T., Belló-Mora, M., & Martinez, D. A. A., (1999). *Satellite Constellation Launch, Deployment, Replacement and End-of-Life Strategies, 1,* 1–33.
19. Crisp, N. H., Smith, K., & Hollingsworth, P., (2015). Launch and deployment of distributed small satellite systems. *Acta Astronautica, 114*, 65–78.
20. Crisp, N., Smith, K., & Hollingsworth, P., (2014). Small satellite launch to LEO: A review of current and future launch systems. *Transactions of the Japan Society for Aeronautical and Space Sciences, Aerospace Technology Japan, 12*(ists29), Tf_39–Tf_47.
21. Del, P. I., Cameron, B. G., & Crawley, E. F., (2019). A technical comparison of three low earth orbit satellite constellation systems to provide global broadband. *Acta Astronautica, 159,* 123–135.
22. Del, R. E., Fantacci, R., & Giambene, G., (1995). Handover and dynamic channel allocation techniques in mobile-cellular networks. *IEEE Transactions on Vehicular Technology, 44*(2), 229–237.
23. Deng, Q., & Simchi-Levi, D., (1993). *Valid Inequalities, Facets and Computational Results for the Capacitated Concentrator Problem* (Vol. 1, pp. 1–23). Technical Report, Department of Industrial Engineering and Operations Research, Columbia University, New York.
24. Díaz-de-Baldasano, M. C., Mateos, F. J., Núñez-Rivas, L. R., & Leo, T. J., (2014). Conceptual design of offshore platform supply vessel based on hybrid diesel generator-fuel cell power plant. *Applied Energy, 116*, 91–100.

25. Falke, A., Hartling, M., Laufer, R., & Roeser, H. P., (2004). Implementation of a simulation environment for software and hardware simulation of power management for a lunar small satellite mission. In: *55th International Astronautical Congress of the International Astronautical Federation, the International Academy of Astronautics, and the International Institute of Space Law* (Vol. 55, pp. IAA-4).
26. Fallon, T., & Oestreich, M., (2015). A constellation of deuterium-labeled silanes as a simple mechanistic probe not requiring absolute configuration determination. *Angewandte Chemie International Edition, 54*(42), 12488–12491.
27. Fossa, C. E., Raines, R. A., Gunsch, G. H., & Temple, M. A., (1998). An overview of the Iridium (R) low earth orbit (LEO) satellite system. In: *Proceedings of the IEEE 1998 National Aerospace and Electronics Conference. NAECON 1998. Celebrating 50 Years (Cat. No. 98CH36185)* (Vol. 1, pp. 152–159). IEEE.
28. Garey, M. R., & Johnson, D. S., (1979). *Computers and Intractability, A Guide to the Theory of NP-Completeness* (Vol. 1, pp. 1–33). Freeman, San Francisco, CA.
29. Gavish, B., & Kalvenes, J., (1995). LEOS-optimal satellite launch policies: The dynamic case. *Operations Research, Under Revision, 1*, 1–30.
30. Gavish, B., & Kalvenes, J., (1996). The impact of satellite altitude on the performance of LEOS based communication systems. *Wireless Networks* (Vol. 1, pp. 1–33).
31. Gavish, B., & Kalvenes, J., (1997a). The impact of inter-satellite communication links on LEOS performance. *Telecommunication Systems* (Vol. 1, pp. 1–31).
32. Gavish, B., & Kalvenes, J., (1997b). LEOS-optimal satellite launch policies: The static case. *Management Science* (Vol. 1, pp. 1–23).
33. Gavish, B., & Neuman, I., (1992). Routing in a network with unreliable components. *IFRP Transactions on Communications, 40*(7), 1248–1258.
34. Gavish, B., & Pirkul, H., (1991). Algorithms for the multi-resource generalized assignment problem. *Management Science, 37,* 695–713.
35. Gavish, B., (1995a). Low earth orbit satellite-based communications systems-some research issues. In: *Proceedings of EURO XIV* (Vol. 1, pp. 15–20). Jerusalem.

36. Gavish, B., (1995b). Telecommunications: A revolution in progress. *Operations Research, 43*(1), 29–32.
37. Gavish, B., (1997a). LEO/MEO systems-global mobile communication systems. *Telecommunication Systems, 1*, 1–24.
38. Gavish, B., (1997b). Routing in low Earth orbit satellite-based communication systems. *Technical Report, Forthcoming, 1*, 1–44.
39. Girard, E., Valerio, G., Ettorre, M., Sauleau, R., & Legay, H., (2015). Physical-optics analysis and design of a beam-forming network coupled to an imaging-system configuration for Ka-band satellite applications. In: *2015 9th European Conference on Antennas and Propagation (EuCAP)* (pp. 1–5). IEEE.
40. Henig, M. I., (1986). The shortest path problem with two objective functions. *European Journal of Operational Research, 25,* 281–291.
41. Henig, M. I., (1994). Efficient interactive methods for a class of multi-attribute shortest path problems. *Management Science, 40*, 891–897.
42. Hinami, K., Farnan, J. M., Meltzer, D. O., & Arora, V. M., (2009). Understanding communication during hospitalist service changes: A mixed-methods study. *Journal of Hospital Medicine: An Official Publication of the Society of Hospital Medicine, 4*(9), 535–540.
43. Hongzheng, C. U. I., & Chao, H., (2011). Satellite constellation configuration design with rapid performance calculation and ordinal optimization. *Chinese Journal of Aeronautics, 24*(5), 631–639.
44. Howard, R. A., (1960). *Dynamic Programming and Markov Processes* (Vol. 1, pp. 1–24). MIT Press, Cambridge, MA.
45. Jakob, P., Shimizu, S., Yoshikawa, S., & Ho, K., (2019). Optimal satellite constellation spare strategy using multi-echelon inventory control. *Journal of Spacecraft and Rockets, 56*(5), 1449–1461.
46. Kaniyil, J., Takei, J., Shimamoto, S., Usui, T., Oka, I., & Kawabata, T., (1992). A global network employing low earth-orbiting satellites. *IEEE Journal on Selected Areas in Communications, 10*(2), 418–427.
47. Lawler, E. L., (1976). *Combinatorial Optimization: Networks and Matroids* (Vol. 1, pp. 1–34). Holt, Rinehart & Winston, New York.
48. Leo, C., & Brown, W., (2000). Slow growth and urban development policy. *Journal of Urban Affairs, 22*(2), 193–213.
49. Leo, C., (1998). Regional growth management regime: The case of Portland, Oregon. *Journal of Urban Affairs, 20*(4), 363–394.

50. Leo, C., Beavis, M. A., Carver, A., & Turner, R., (1998). Is urban sprawl back on the political agenda? Local growth control, regional growth management, and politics. *Urban Affairs Review, 34*(2), 179–211.
51. Li, H. N., Li, J. S., & Jiao, W. H., (2010). Analyzing perturbation motion and studying configuration maintenance strategy for compass-M navigation constellation. *J. Astronaut, 31*(7), 1756–1761.
52. Li, V., & Silvester, J., (1986). Performance analysis of networks with unreliable components. *IEEE Transactions on Communications, 32*(10), 1105–1110.
53. Lienig, J., Jerke, G., & Adler, T., (2002). Electromigration avoidance in analog circuits: Two methodologies for current-driven routing. In: *Proceedings of ASP-DAC/VLSI Design 2002. 7th Asia and South Pacific Design Automation Conference and 15th International Conference on VLSI Design* (Vol. 1, pp. 372–378). IEEE.
54. Luo, Y., & Ansari, N., (2005). Bandwidth allocation for multiservice access on EPONs. *IEEE Communications Magazine, 43*(2), S16–S21.
55. Maier, M. W., Wendoloski, E., Houston, D., & Wilson, J., (2018). Launch and production schedule modeling for sustained earth observation constellations. In: *2018 IEEE Aerospace Conference* (Vol. 1, pp. 1–16). IEEE.
56. Maral, G., De Ridder, J. J., Evans, B. G., & Richharia, M., (1991). Low earth orbit satellite systems for communications. *International Journal of Satellite Communications, 9*, 209–225.
57. Markowic, Z., & Hope, W., (1992). Small, low earth orbit communication satellites: An evaluation. In: *IREECON '92, Australia's Electronics Convention Proceedings* (Vol. 1, pp. 178–181). IREE, Edgecliff, NSW.
58. McDowell, J. C., (2020). The low earth orbit satellite population and impacts of the space X star link constellation. *The Astrophysical Journal Letters, 892*(2), L36.
59. Mikkonen, M., Va, S., Ikonen, V., & Heikkila, M. O., (2002). User and concept studies as tools in developing mobile communication services for the elderly. *Personal and Ubiquitous Computing, 6*(2), 113–124.
60. Mostacciuolo, E., Iannelli, L., Sagnelli, S., Vasca, F., Luisi, R., & Stanzione, V., (2018). Modeling and power management of a LEO small satellite electrical power system. In: *2018 European Control Conference (ECC)* (Vol. 1, pp. 2738–2743). IEEE.

61. Newman, L., Biedrzycki, K., & Baum, F., (2012). Digital technology use among disadvantaged Australians: Implications for equitable consumer participation in digitally-mediated communication and information exchange with health services. *Australian Health Review, 36*(2), 125–129.
62. Perrotta, G., (1991). A comparison between low circular and elliptical inclined orbits for small satellites communication applications. In: *Proceedings Second European Conference on Satellite Communications* (pp. 431–434). Vol. ESA SP-322, ESA, Liege.
63. Powell, T. M., Lewis, C. V., Curchitser, E. N., Haidvogel, D. B., Hermann, A. J., & Dobbins, E. L., (2006). Results from a three-dimensional, nested biological-physical model of the California current system and comparisons with statistics from satellite imagery. *Journal of Geophysical Research: Oceans, 111*(C7), 1–33.
64. Pratt, S. R., Raines, R. A., Fossa, C. E., & Temple, M. A., (1999). An operational and performance overview of the Iridium low earth orbit satellite system. *IEEE Communications Surveys, 2*(2), 2–10.
65. Radhakrishnan, R., Edmonson, W. W., Afghah, F., Rodriguez-Osorio, R. M., Pinto, F., & Burleigh, S. C., (2016). Survey of inter-satellite communication for small satellite systems: Physical layer to network layer view. *IEEE Communications Surveys and Tutorials, 18*(4), 2442–2473.
66. Radhakrishnan, R., Edmonson, W. W., Afghah, F., Rodriguez-Osorio, R. M., Pinto, F., & Burleigh, S. C., (2016). Survey of inter-satellite communication for small satellite systems: Physical layer to network layer view. *IEEE Communications Surveys and Tutorials, 18*(4), 2442–2473.
67. Rider, L., (1985). Optimized polar orbit constellations for redundant earth coverage. *Journal of the Astronautical Sciences, 33*(2), 147–161.
68. Rider, L., (1986). Analytic design of satellite constellations for zonal coverage using inclined circular orbits. *Journal of the Astronautical Sciences, 34*(1), 31–64.
69. Sanso, B., & Soumis, F., (1991). Communication and transportation network reliability using routing models. *IEEE Transactions on Reliability, 40*, 29–38.
70. Shah, S. H., Chen, K., & Nahrstedt, K., (2005). Dynamic bandwidth management in single-hop ad hoc wireless networks. *Mobile Networks and Applications, 10*(1), 199–217.

71. Sheriff, R. E., & Gardiner, J. G., (1993). The applicability of LEO satellites to 3rd generation networks. In: *Fourth IEEE Conference on Telecommunications, Conf. Publ. No. 371* (pp. 296–300). *IEEE,* London.
72. Shi, R., Liu, L., Long, T., Wu, Y., & Wang, G. G., (2018). Multidisciplinary modeling and surrogate assisted optimization for satellite constellation systems. *Structural and Multidisciplinary Optimization, 58*(5), 2173–2188.
73. Shogan, A., (1976). Sequential bounding of the reliability of a stochastic network. *Operations Research 34*(6), 1027–1044.
74. Shogan, A., (1978). A decomposition algorithm for network reliability analysis. *Networks, 8*, 231–251.
75. Smith, E. A., Cooper, H. J., Xiang, X., Mugnai, A., & Tripoli, G. J., (1992). Foundations for statistical-physical precipitation retrieval from passive microwave satellite measurements. Part I: Brightness-temperature properties of a time-dependent cloud-radiation model. *Journal of Applied Meteorology and Climatology, 31*(6), 506–531.
76. Wang, C., Zhao, Q., Guo, J., Liu, J., & Chen, G., (2019). The contribution of inter-satellite links to BDS-3 orbit determination: Model refinement and comparisons. *Navigation, Journal of the Institute of Navigation, 66*(1), 71–82.
77. Ware, R., Exner, M., Feng, D., Gorbunov, M., Hardy, K., Herman, B., & Trenberth, K., (1996). GPS sounding of the atmosphere from low earth orbit: Preliminary results. *Bulletin of the American Meteorological Society, 77*(1), 19–40.
78. Werner, M., Delucchi, C., Vogel, H. J., Maral, G., & De Ridder, J. J., (1997). ATM-based routing in LEO/MEO satellite networks with inter-satellite links. *IEEE Journal on Selected Areas in Communications, 15*(1), 69–82.
79. Wolsey, L. A., (1989). Sub modularity and valid inequalities in capacitated fixed charge networks. *Operations Research Letters, 8*, 119–124.
80. Wu, J., Cheng, Y. J., & Fan, Y., (2014). A wideband high-gain high-efficiency hybrid integrated plate array antenna for V-band inter-satellite links. *IEEE Transactions on Antennas and Propagation, 63*(4), 1225–1233.

81. Zaitchik, B. F., Rodell, M., & Olivera, F., (2010). Evaluation of the global land data assimilation system using global river discharge data and a source-to-sink routing scheme. *Water Resources Research, 46*(6), 1–24.

Chapter

4

MEDIUM EARTH ORBIT SATELLITES

CONTENT

4.1. Introduction 100
4.2. Satellite Constellation Architecture Design Tradeoffs 100
4.3. Regional Layout and Gateways 103
4.4. Low Latency Advantages For Interactive Applications 103
4.5. Impact of High Latency on File Transfer and Web-Based C^2 Applications Over Tcp/Ip 105
4.6. High Data Rate Advantages 107
4.7. Implications for Ground Terminal Equipment 107
References 109

4.1. INTRODUCTION

To underserved locations, the medium earth orbit (MEO) satellites provide fiber-like low latency. In this chapter, we will describe the design of an effective and economical low latency MEO constellation by the usage of high-power dirigible Ka-band spot beams. In case those applications are less delay-sensitive, high-speed ISR data could be transferred from single tracking antennas deliver from sensor stages to the processing center and the quick distribution of managed intelligence information back to forward control posts and smaller groups (Bergstra and Middelburg, 2003; Skurowski *et al.,* 2010).

On Earth, MEO satellites bring fiber-like low potential to underserved locations. This chapter explains the design of an effective, economical low latency MEO constellation. Through the high-power dirigible, Ka-band spot beams give high-speed data (HSD) transfer from inspection and investigation sensors to intelligence analysis and managing centers and the swift distribution of managed intelligence information back to forward command posts and smaller groups (Lyras et al., 2019; Skoulidou et al., 2019). A smooth error-free satellite-to-satellite transfer by applying dual receiver modems and tracking antennas has been utilized to provide continuous real-time execution for these cooperative time-sensitive applications (Sing and Soh, 2004; Blumenthal, 2013).

4.2. SATELLITE CONSTELLATION ARCHITECTURE DESIGN TRADEOFFS

At an altitude of 35,786 km over sea level in an Equatorial orbit GEO (Geosynchronous Earth Orbit) satellites have the benefit that the satellites rotate around the Earth at the same speed that the Earth goes on its axis and so the satellites stay in a comparatively fixed position on a precise location on the Earth all the time. This makes things easier than the ground stations needed to connect with the satellites because they can stay pointed at a similar location in the sky every time (Sivchenko et al., 2004; Border *et al*., 2020).

Though, the very high round trip dormancy of 500–600 msec effects the action of several communications applications. Some non-geosynchronous orbit telecommunications systems have been installed and built (Globalstar and Iridium), and a few other systems have been planned (ICO, Celestri, Teledesic). The major benefit of employing an orbit lower than GEO is that round trip invisibility can be substantially decreased, which enhances the action of certain applications (Etsi, 2006; Huang *et al*., 2010). Systems like Globalstar and Iridium have been utilized into LEO (low earth orbit) and are intended to provide mainly voice and very decreased data rate services. Based on the markets, these systems produced a few design decisions that they were intending to serve. To assist in lower latency and simplify the ground infrastructure, these systems utilize inter-satellite communications links. This permits traffic to go from one location on the surface of Earth to another without several up and down trips among ground stations and satellites. This factor entails a necessity on the ground systems and to satellite be capable to track and point narrow down communications beams precisely at objects that are all moving vigorously comparative to each other (Rossi, 2008; Lücking et al., 2012).

The O3b Networks has developed an MEO constellation. O3b means "the other 3 billion," and their system is devised to give very high data rate low potential fiber-like services to underserved markets in the evolving world. Tradeoffs among orbital altitude and coverage area showed that a primarily MEO constellation in a circular Equatorial would offer adjacent coverage to territories inside the underserved parts of the evolving world (Corazza and Vatalaro, 1994). A primary constellation of 8 satellites, at this equatorial MEO altitude, provides continuous service to each part of the Earth within 45° of the Equator. This is the region of the Earth that covers several main "hot spots." This constellation can also give great bandwidth services for disaster relief, emergency responders, and fiber restoral (Ferringer *et al*., 2007). In fixed orbit locations, the GEO satellites are typically placed. In several cases, their satellite antenna exuding patterns have been designed to just give coverage above the landmasses beneath them. As the MEO satellites in their orbit are moving, they give similar coverage above the oceans as they act over the landmasses inside the +/–45° latitudes coverage area. The round-trip latency is usually below 130 msec at an 8000 km orbital altitude and is promised to always be < 150 msec inside the coverage region being operated. This is similar to the fiber routes of long haul and about 4 times < GEO satellite round trip latency (Whittecar and Ferringer, 2014; Paek *et al*., 2018).

The O3b networks MEO system has yielded numerous other design tradeoffs centered on the types of service that would be proposed in these underserved emerging world markets. Figure 4.1 displays the MEO satellite gathering coverage area from an Equatorial orbit together with examples of customer beams (open circles) and Gateway beams (filled circles).

Figure 4.1. *Medium earth orbit satellite constellation displaying customer and gateway spot beams.*

Source: https://spaceflight101.com/spacecraft/o3b/.

Getting the satellites nearer to Earth showed that signals among ground stations and satellites gain 13 dB less path loss. At the surface of the Earth, MEO satellites can give a similar flux density as GEO satellites along with 13 dB less EIRP (equivalent isotropic radiated power) after the satellites and can attain comparable receive sensitivity as GEO satellites with minor aperture antennas. Working in the industrial Ka-band along with shorter wavelengths vs. Ku or C band systems also aids to decrease the ground station and satellite antenna aperture size whereas providing similar radiated power levels and receive sensitivity (Bell *et al.,* 2000).

Because of the lower power needs, it permitted the satellites to be slighter and to weigh < 700 kg, decreasing satellite cost. In order to deploy eight satellites, two launches will be necessary into the 8062 km MEO. The system is extremely accessible, and extra satellites can be improved to the assemblage over time to add further capacity.

The satellite communications payload is involved in another set of design tradeoffs. The decision of O3b's to function in the commercial Ka-band permitted it to get access to a downlink frequency range of 17.8–19.3 GHz and 1.5 GHz of the spectrum along with an uplink frequency range of 27.6–29.1 GHz. As a modest RF bent pipe, the satellite payload has been used along with no committed baseband processing. The payload contains 10 customer beams and two Gateway beams. With every, the 10 customer beams contain 216 MHz of bandwidth in every direction. Usage of the left hand and right-hand circular polarization allows frequency reuse. Every Gateway beam and every customer beam is applied by a small dirigible gimbaled parabolic tracking antenna for a whole of 12 antennas (10 for customer beams and 2 for Gateway beams) on every satellite.

4.3. REGIONAL LAYOUT AND GATEWAYS

In the assemblage being developed by O3b, the eight satellites having an orbital period of about 288 minutes and are present at an orbit of about 8062 km. Each one of which revolves in the same direction in which Earth is revolving around its axis, and they revolve five times per day around the Earth. Satellites remain for 45 minutes over each region of the Earth. After 45 minutes, the satellite from that region moves to the next region of Earth towards the east to compensate that specific region; another satellite in the constellation moves towards the west (David and Bunn, 1988; Yang *et al*., 2019). In each region, this needs a handover to the next moving satellite from the setting satellite. This handover in each region from one satellite to another is not quick; it takes some time and before the completion of handover in the previous region. O3b has permitted a little amount of the Handover Interval, overlap time to finish this. To calculate for this extra time, the quantity of this extra time, for the primary eight satellite constellation active region numbers reduced to seven service regions each one of these seven service regions is joined by a pathway that connects it to the terrestrial fiber infrastructure. For large flexibility in between the regional scope area, the two pathway beams on one of each satellite can be separately pointed to one of the two different pathway locations (Li and Li, 2009; Wang and Ducruet, 2012).

4.4. LOW LATENCY ADVANTAGES FOR INTERACTIVE APPLICATIONS

Interactive applications with less latency perform better. Examples are video conferencing and conversational voice. Some business applications are web access, enterprise resource planning systems, web content download, video streaming, and interactive gaming. HTTPS protocols are used in video streaming over TCP.

4.4.1. Interactive Conversational Voice and Video Conferencing

Around 250–300 msec delay in the round-trip results in an unnatural quality of voice conversations and also decreases their instructiveness. When one speaker finishes the talk, next comes then. E-model or Mean Opinion

Score is used to calculate the quality of voice calls which is defined by ITU Recommendation G.107. For the single-way mouth-to-ear transmission delay, ITU Recommendation G.114 used which is based on the E model. In Figure 4.2, the graph which uses ITU Recommendation G.114 indicates E-model's degradation of R rating shows that when single-way transmission delay reaches above the level of 275 msec, users become unhappy or dissatisfied (Sat and Wah, 2007; Saidi et al., 2016).

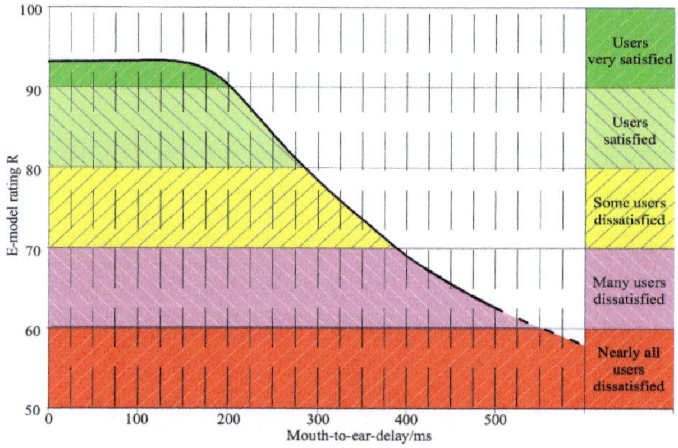

Figure 4.2. *E-model R rating performance vs. one-way delay via ITU recommendation G.*

Source: https://ieeexplore.ieee.org/abstract/document/6735634.

For videoconferences, the same type of phenomenon occurs. In addition to the decrease of voice conversation interactivity, there is a loss in facial queues due to high latency as when the talk is finished by the speaker. Videoconferences and Voice over greater latency links mostly perceive more time which is less user Voice and videoconferences over high latency links generally take more time resulting in lower user productivity (Daly-Jones et al., 1998).

4.4.2. Enterprise Resource Planning and Distributed C^2 Systems

Oracle and SAP type of Enterprise Resource Planning systems support remote access. In the case of each transaction for these applications, more handshakes are observed between server and remote access client. One problem is with 500–600 msec use of satellite (GEO) round trip latency for every several round-trip delays, handshake, and response time for client-server becomes very less which reduces the work productivity, remote staff. These involve modifications at the application level to the amount of information that is used or sent through the link or remote terminal servers at the remote site used to implement a part of the enterprise application. For example, in order to limit the sent information at low latency throughput links or high latency by using an Oracle White Paper. High delay satellite links contain many problems faced by civilian ERP. User productivity and application responsiveness are improved by using Lower latency, and it is required to develop custom software workarounds (Bradley, 2008; Lee and Wang, 2019).

4.5. IMPACT OF HIGH LATENCY ON FILE TRANSFER AND WEB-BASED C² APPLICATIONS OVER TCP/IP

The TCP (transmission control protocol) is intended to yield an end-to-end error-free, sequenced supply of packets for specific applications or users over an inaccurate IP (internet protocol) infrastructure of networks interlocked by IP routers. To accomplish this, the TCP executes not many services that give congestion prevention, end-to-end management of flow, packet supply acknowledgment and retransmission, and the reasonable usage of the basic shared transport networks. The action of ` on great bandwidth high latency Geosynchronous Earth Orbit satellite links has been examined widely by Partridge and Shepard (1997); and Allman et al. (1999). The local storing on the other end of the satellite link can also participate in enhanced performance for static content. Though methods like PEP (performance-enhancing proxies) are explained by RFC 3135 function well for satellite links along very low BER, their performance decreases as bit error rates increases (Pereira et al., 2011).

The graphs demonstrate that lower latency enhances output for those applications that work over TCP. Protocols like local and PEP caching methods can more improve link output even for lower latency links.

In the case of military systems that may include transmission of categorized data above end-to-end encrypted links, the advantages of local and PEP caching cannot be achieved because of the incompetence to gain access to the TCP headers and content of the encrypted data stream. Along with higher per-user output, lower latency also enhances response time for collaborative data gain access from an inaccessible location, particularly for dynamic data that cannot be simply cached at the isolated end of the link. This indicates greater productivity because of lower wait times for data, particularly if retransmissions are needed around links with high BET (Figures 4.3 and 4.4) (Butts *et al.*, 1999).

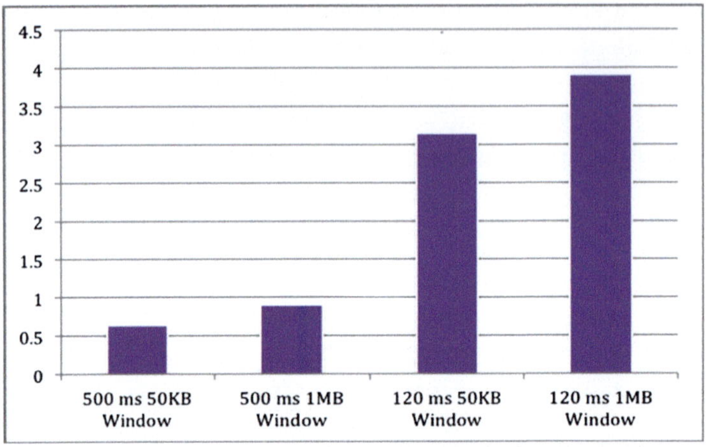

Figure 4.3. *TCP alone-amount in Mbps; 1500-byte packets, 50 Kbyte, and 1 Mbyte TCP window sizes, and 500 msec and 120 msec link round trip.*

Source: https://www.semanticscholar.org/paper/Medium-Earth-Orbit-Ka-Band-Satellite-Communications-Blumenthal/0af3f86ee32504bc030c6513c130 3ca1a1b08361.

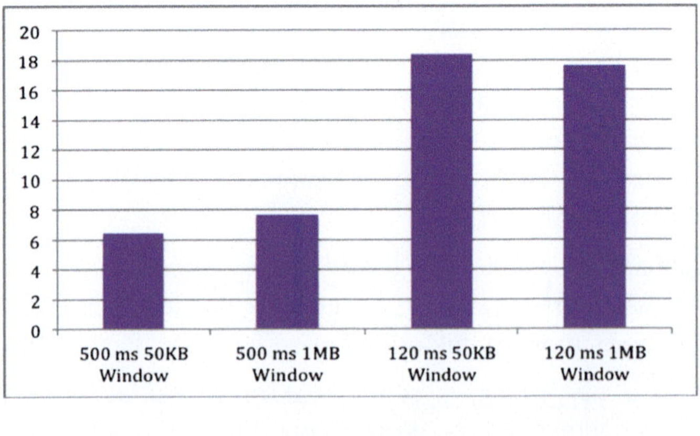

Figure 4.4. *TCP with PEP enabled-throughput in Mbps; 1500-byte packets, 50 Kbyte and 1 Mbyte TCP window sizes, and 500 msec and 120 msec link round trip latencies.*

Source: https://www.researchgate.net/publication/271555403_Medium_ Earth_Orbit_Ka_Band_Satellite_Communications_System.

4.6. HIGH DATA RATE ADVANTAGES

In order to transfer high-resolution multispectral imagery back from areas of discord to intelligence processing and aggregation centers, ISR (intelligence, surveillance, and reconnaissance) missions need high data rates. Around 100 Mbps of data, the high-capacity beams can recover via a small ISR platform fitted out with the right size tracking antenna. In order to conceal an area of interest, dirigible spot beams can be swiftly moved and can follow a moving ISR platform.

Portable, quickly deployable terminal equipment can be located at main locations around the world to assist disaster relief efforts and emergency responders. Inside an hour dirigible, spot beams can be moved to specific locations (Millington and Isted, 1950; Morea and Rival, 2010). Due to the very low latency and high bandwidth available, the services of fiber restoral are being offered by the MEOs satellites can get in touch with the data rate and execution of the fiber connections (Videler and Weihs, 1982; Waddington, 2009).

4.7. IMPLICATIONS FOR GROUND TERMINAL EQUIPMENT

In order to accomplish high output on the links, in the satellites, very directional gimbaled tracking antennas are applied to establish spot beams that are kept directed at a fixed location on the Earth, but the satellite is moving. With the help of a single tracking antenna or with double-tracking antennas, ground terminals can be utilized to give two different modes of operation (Zanetti et al., 2013).

The execution of single-antenna terminals a break before the formation of Satellite-to-Satellite transfer. By the end of the 45-minute satellite pass, the antenna of the ground terminal shifts back to develop the next rising satellite. In the data stream, these outcomes in a scheduled break, but the tracking antenna is tracking and locking on the next satellite, moving back, permitting its dial-up connection to attain carrier sync and yield good frames, and then allowing any encryption and router devices to re-sync if essential. In hundreds of msec, electronically guided flat panel array antennas can alter beam pointing, and for the single antenna, mode operations are being examined. There may be specific applications, for example, store and forward exploration data recovery that can protect data for a few minutes on the small platform and operate with a single antenna.

If constant data transfer is mandatory, the dual antennas can be installed to implement seamless make before break satellite-to-Satellite transfer. Throughout the 45-minute pass, one antenna is following the satellite. The 2nd antenna is on standby to develop the next rising satellite when it occurs in the region. When the 2nd antenna tracking the rising satellite and is locked on, the modulators at each end of the link transfer a similar signal above both antennas and across the setting and rising satellites for a short instant. At every end of the link, the signals above the 2nd antenna and rising satellite path are obtained through a separate 2nd decoder/demodulator in the modems. When the 2nd decoder/demodulator at every link has locked on the new carrier and is delivering good frames, the modulators cease transferring over the 1st antenna and the 1st decoders/demodulators. The 2nd antenna that is tracking the rising satellite has now become the main antenna for the following 45-minute pass. Through the pass, the 1st antenna changes back to expect for the next increasing satellite, and the handover procedure repeats. A block diagram of a dual antenna terminal is being shown in Figure 4.5 (Slater and Niemi, 2003).

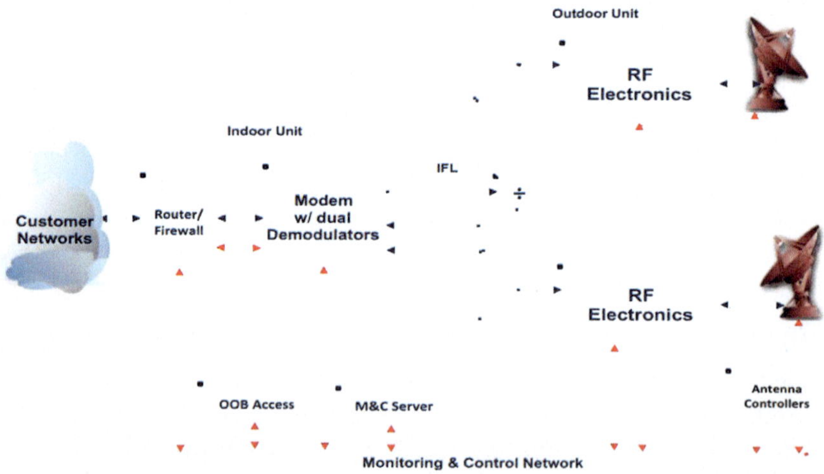

Figure 4.5. *Dual antenna of ground terminal.*

Source: https://ieeexplore.ieee.org/document/6735634/.

REFERENCES

1. Allman, M., Glover, D., & Sanchez, L., (1999). *Enhancing TCP Over Satellite Channels Using Standard Mechanisms, 1,* 1–24.
2. Bell, D. J., Cesarone, R., Ely, T., Edwards, C., & Townes, S., (2000). Mars network: A mars orbiting communications and navigation satellite constellation. In: *2000 IEEE Aerospace Conference. Proceedings (Cat. No. 00TH8484)* (Vol. 7, pp. 75–88). IEEE.
3. Bergstra, J. A., & Middelburg, C. A., (2003). *ITU-T Recommendation G. 107: The E-Model, a Computational Model for Use in Transmission Planning, 1,* 1–33.
4. Blumenthal, S. H., (2013). Medium earth orbit Ka-band satellite communications system. In: *MILCOM 2013–2013 IEEE Military Communications Conference* (Vol. 1, pp. 273–277). IEEE.
5. Border, J., Shah, B., Su, C. J., & Torres, R., (2020). Evaluating QUIC's performance against performance-enhancing proxy over satellite link. In: *2020 IFIP Networking Conference (Networking)* (Vol. 1, pp. 755–760). IEEE.
6. Bradley, J., (2008). Management-based critical success factors in the implementation of enterprise resource planning systems. *International Journal of Accounting Information Systems, 9*(3), 175–200.
7. Butts, N. P., Bharadwaj, V. G., & Baras, J. S., (1999). Internet service via broadband satellite networks. In: *Multimedia Systems and Applications* (Vol. 3528, pp. 169–180). International Society for Optics and Photonics.
8. Corazza, G. E., & Vatalaro, F., (1994). A statistical model for land mobile satellite channels and its application to nongeostationary orbit systems. *IEEE Transactions on Vehicular Technology, 43*(3), 738–742.
9. Daly-Jones, O., Monk, A., & Watts, L., (1998). Some advantages of video conferencing over high-quality audio conferencing: Fluency and awareness of attentional focus. *International Journal of Human-Computer Studies, 49*(1), 21–58.
10. David, P. A., & Bunn, J. A., (1988). The economics of gateway technologies and network evolution: Lessons from electricity supply history. *Information Economics and Policy, 3*(2), 165–202.
11. Etsi, E. N., (2006). 302 307:" Digital video broadcasting (DVB). *Second Generation Framing Structure, Channel Coding and Modulation Systems for Broadcasting, Interactive Services, News Gathering and Other Broadband Satellite Applications, 1,* 2.

12. Ferringer, M. P., Clifton, R. S., & Thompson, T. G., (2007). Efficient and accurate evolutionary multi-objective optimization paradigms for satellite constellation design. *Journal of Spacecraft and Rockets, 44*(3), 682–691.
13. Huang, L., Qiu, X., Hu, D., & Ding, C., (2010). Focusing of medium-earth-orbit SAR with advanced nonlinear chirp scaling algorithm. *IEEE Transactions on Geoscience and Remote Sensing, 49*(1), 500–508.
14. Lee, H. Y., & Wang, N. J., (2019). Cloud-based enterprise resource planning with elastic model-view-controller architecture for internet realization. *Computer Standards and Interfaces, 64*, 11–23.
15. Li, X., & Li, Z., (2009). A new programming model for placement and selection of dynamic gateways based on MANET accessing internet. *Acta Electronica Sinica, 4*, 1–33.
16. Lücking, C., Colombo, C., & McInnes, C. R., (2012). A passive satellite deorbiting strategy for medium earth orbit using solar radiation pressure and the J2 effect. *Acta Astronautica, 77*, 197–206.
17. Lyras, N. K., Efrem, C. N., Kourogiorgas, C. I., & Panagopoulos, A. D., (2019). Medium earth orbit optical satellite communication networks: Ground terminals selection optimization based on the cloud-free line - of - sight statistics. *International Journal of Satellite Communications and Networking, 37*(4), 370–384.
18. Millington, G., & Isted, G. A., (1950). Ground-wave propagation over an inhomogeneous smooth earth. Part 2: Experimental evidence and practical implications. *Proceedings of the IEE-Part III: Radio and Communication Engineering, 97*(48), 209–221.
19. Morea, A., & Rival, O., (2010). Advantages of elasticity versus fixed data-rate schemes for restorable optical networks. In: *36th European Conference and Exhibition on Optical Communication* (Vol. 36, pp. 1–3). IEEE.
20. Paek, S. W., Kronig, L. G., Ivanov, A. B., & De Weck, O. L., (2018). Satellite constellation design algorithm for remote sensing of diurnal cycles phenomena. *Advances in Space Research, 62*(9), 2529–2550.
21. Partridge, C., & Shepard, T. J., (1997). TCP/IP performance over satellite links. *IEEE Network, 11*(5), 44–49.
22. Pereira, P. R., Casaca, A., Rodrigues, J. J., Soares, V. N., Triay, J., & Cervelló-Pastor, C., (2011). From delay-tolerant networks to vehicular delay-tolerant networks. *IEEE Communications Surveys and Tutorials, 14*(4), 1166–1182.

23. Rossi, A., (2008). Resonant dynamics of medium earth orbits: Space debris issues. *Celestial Mechanics and Dynamical Astronomy, 100*(4), 267–286.
24. Saidi, I., Zhang, L., Barriac, V., & Deforges, O., (2016). Interactive vs. non-interactive subjective evaluation of IP network impairments on audiovisual quality in video conferencing context. In: *2016 Eighth International Conference on Quality of Multimedia Experience (QoMEX)* (Vol. 1, pp. 1–6). IEEE.
25. Sat, B., & Wah, B. W., (2007). Playout scheduling and loss-concealments in VoIP for optimizing conversational voice communication quality. In: *Proceedings of the 15th ACM International Conference on Multimedia* (Vol. 1, pp. 137–146).
26. Sing, J., & Soh, B., (2004). TCP performance over geostationary satellite links: Problems and solutions. In: *Proceedings. 2004 12th IEEE International Conference on Networks (ICON 2004) (IEEE Cat. No. 04EX955)* (Vol. 1, pp. 14–18). IEEE.
27. Sivchenko, D., Xu, B., Zimmermann, G., & Hischke, S., (2004). Internet traffic performance in high-speed trains. In: *HET-NETs' 04: Second Int'l Working Conf.: Performance Modeling and Evaluation of Heterogeneous Networks* (Vol. 1, pp. 26–31).
28. Skoulidou, D. K., Rosengren, A. J., Tsiganis, K., & Voyatzis, G., (2019). Medium earth orbit dynamical survey and its use in passive debris removal. *Advances in Space Research, 63*(11), 3646–3674.
29. Skurowski, P., Wójcicki, R., & Jerzak, Z., (2010). Evaluation of IP transmission jitter estimators using one-way active measurement protocol (OWAMP). In: *International Conference on Computer Networks* (Vol. 1, pp. 153–162). Springer, Berlin, Heidelberg.
30. Slater, L., & Niemi, T. M., (2003). Ground-penetrating radar investigation of active faults along the dead sea transform and implications for seismic hazards within the city of Aqaba, Jordan. *Tectonophysics, 368*(1–4), 33–50.
31. Videler, J. J., & Weihs, D., (1982). Energetic advantages of burst-and-coast swimming of fish at high speeds. *Journal of Experimental Biology, 97*(1), 169–178.
32. Waddington, L., (2009). Breaking new ground: The implications of ratification of the UN convention on the rights of persons with

disabilities for the European community. In: *The UN Convention on the Rights of Persons with Disabilities* (Vol. 1, pp. 111–140). Brill Nijhoff.

33. Wang, C., & Ducruet, C., (2012). New port development and global city making: Emergence of the Shanghai-Yangshan multilayered gateway hub. *Journal of Transport Geography, 25*, 58–69.
34. Whittecar, W. R., & Ferringer, M. P., (2014). Global coverage constellation design exploration using evolutionary algorithms. In: *AIAA/AAS Astrodynamics Specialist Conference* (Vol. 1, p. 4159).
35. Yang, Z., Guo, L., & Lian, F., (2019). Port integration in a region with multiport gateways in the context of industrial transformation and upgrading of the port. *Transportation Research Part E: Logistics and Transportation Review, 122*, 231–246.
36. Zanetti, E. M., Bignardi, C., Franceschini, G., & Audenino, A. L., (2013). Amateur football pitches: Mechanical properties of the natural ground and of different artificial turf infills and their biomechanical implications. *Journal of Sports Sciences, 31*(7), 767–778.

Chapter

5

GEOSTATIONARY SATELLITES AND THEIR CONSTELLATIONS

CONTENT

5.1. Introduction..114

5.2. Development of Geostationary Satellites...116

5.3. Satellite Constellations: Many Architectures and Many Problems....119

References..121

5.1. INTRODUCTION

A satellite with an orbital period the same as the Earth's rotation period in geosynchronous orbits is called a geosynchronous satellite. The geostationary satellite is a specific case of geosynchronous satellite, which contains a circular geosynchronous orbit-a geostationary orbit (GEO) directly over the Earth's equator. The Tundra elliptical orbits an additional kind of geosynchronous orbit employed by satellites (Picone *et al.,* 2005; Johnston, 2013).

Geostationary satellites have the exceptional characteristics of staying constantly fixed in precisely the similar position in the sky as seen from any permanent location on Earth, indicating that ground-based antennas do not require to trace them but can stay fixed in one direction. These satellites are frequently employed for communication objectives; a communication network centered on communication over geosynchronous satellites is called a geosynchronous network (Clarke, 1966; Leese *et al*., 1971).

5.1.1. Definition

The term *geosynchronous* is used for the satellite's orbital period, which permits it to be along with the rotation of the Earth (geo "means rotation of earth"). With the necessity of this orbital period, to be *geostationary* too, in an orbit, the satellite should be positioned that puts it in the locality around the equator. For communications satellites, the distinct case of a GEO is the very ordinary type of orbit (Zhang *et al.,* 2012; Del Portillo *et al*., 2019).

If a geosynchronous satellite's orbit is not precisely associated with the Earth's equator, the orbit is taken as a favored orbit. Around a stable point, it will seem to oscillate every day (Draim *et al*., 2000).

5.1.2. Application

There are nearly 446 operational geosynchronous satellites; few of them are not operative.

At every place where the satellite is evident, an observer will constantly see it in precisely a similar spot in the sky, unlike planets and stars that move continuously (Kodheli et al., 2017).

On the Earth, transmitting and receiving antennas do not require to track such a satellite. In a place, these antennas can be installed and are economical than tracking antennas. These satellites have developed television broadcasting and weather forecasting, global communications, and contain a few important intelligence and defense applications (Sánchez *et al.,* 2017).

The outcome of their high altitude is one drawback of geostationary satellites: in order to reach and come back from the satellite, radio signals take roughly 0.25 of a second, causing a small but considerable signal delay. There are a few commercial satellite data protocols that are intended to substitute TCP/IP connections above long-delay satellite links—these are promoted as being a limited solution to the bad performance of resident TCP above satellite links. TCP assumes that all damage is because of congestion, not inaccuracies, and investigations link capacity with its "slow-start" algorithm, which just delivers packets when it is known that previous packets have been collected. By employing a geostationary satellite, slow start is extremely slow across a path (Wang et al., 2014; Knepp et al., 2015).

The advantages of geostationary satellites are given (Ineichen and Perez, 1999; Liu and Lin, 2004):

- Satellite always lies in a similar position;
- The tracking of the satellite through its earth stations is easy;
- Find high sequential resolution data.

The inadequate geographical coverage is also a disadvantage of geostationary satellites because ground stations at greater than nearly 60° latitude have trouble at low elevations, dependably receiving signals. Across the greatest amount of atmosphere, the signals would require to pass, and could even be stopped by land buildings, vegetation, or topography. A useful solution was established in the USSR, for this challenge with the formation of specific Molniya path satellite networks along with elliptical orbits. For the Sirius Radio satellites like elliptical orbits are employed (Malik et al., 1991).

5.1.3. History

In 1928, this concept was first proposed by Herman Potočnik, and in 1945 it was popularized by the science fiction author Arthur C. Clarke in a paper in *Wireless World.* Functioning earlier to the introduction of solid-state electronics, Clarke proposed a trinity of large, occupied space stations positioned in a triangle across the planet. There are several uncrewed satellites and usually no bigger than an automobile (Yang *et al.,* 2016).

The first operational geosynchronous satellite named "Syncom 2" was invented by the "father of the geosynchronous satellite," Harold Rosen, an engineer at Hughes Aircraft Company. On July 26, 1963, it was introduced on the Delta rocket B booster from Cape Canaveral.

On August 19, 1964, the initial geostationary communication satellite named Syncom 3 was launched, with a Delta D launch vehicle from Cape Canaveral. The first American domestic and commercially geostationary communications satellite called Westar 1 was launched by Western Union and NASA on April 13, 1974.

Since Echo 1, the satellites have been applied in telecommunications, on August 12, a 26.5-inch magnesium sphere was introduced by a Thor Delta rocket, 1560 bounded a recorded message spread front Goldstone, California that was taken by the Belt telephone laboratory an N. J. Holmdel, Echo I enthused a good deal of interest in the advancement of active communication which guides AT&T (American Telephone and Telegraph Company) to shape Telstar, on July 10, 1962, it was launched along with a microwave transmitter and receiver Telstar was an active satellite. It was the primary satellite to send out live television and exchanges around the Atlantic (Zhu *et al.*, 2013).

In 1947 Geostationary satellites were introduced by Arthur (1917), he was a British astronomer and physicist as a means to depend on radio signals from one part to another that is outside the line of prospect GEOs are orbits filled by communications satellites which stay at secured points in the sky comparative to observers on the ground. Around its polar axis during one sidereal day, the earth rotation of Earth occurs only once this period describes the average orbit radius of" 42155 km from Kepler's third law this value is originated. From the orbit radius the Earth's radius (6370 km) is deducted which regulates the orbit over the Earth to be 3ñ 78a km, about the shape of the orbit this explanation does not say anything, or the alignment of the orbit plane regarding the plane of the equator the orbit can be extremely oblique, or it can be tending regarding the plane of the equator. In this case, it is however be synchronic with the Earth's rotation. The GEO is a preferred class of geosynchronous orbit (Rao *et al.*, 1990).

5.2. DEVELOPMENT OF GEOSTATIONARY SATELLITES

In the sky, a satellite traveling in a stationary orbit stays at a fixed point every time. For' radio communications this is suitable because it permits the usage of static antennas on the ground.

There are three criteria that an orbit must meet to be geostationary:

- An orbit should lie in the Earth's equatorial plane;
- The orbit should be geosynchronous; and
- The orbit should be a circle.

Specific satellites inside the orbit are detected by the longitudinal position west or east of the prime meridian with the following whole GEOs comply:

Peter	Metric Units
Orbit circumference	264865 km
Height above equator	35785 km
Arc length per degree	736 km
Ayer age orbit radius	42155 km
Orbital velocity	1,1066 km/h
Total	XXXX

In order to get the know-how of these criteria. Consider the outcome if the orbit collapses to encounter them then the orbit is non-geosynchronous, the satellite does not move therefore, from the perspective of a viewer on Earth, in continuous motion, the satellite appears, and it periodically vanishes under the horizon. At a constant velocity, the satellite does not move about if the orbit is not a circle. As an alternative, at a rate of two cycles per sidereal day, it appears to oscillate east-and-west. If the orbit does not recline in the equatorial plane, the satellite does not remain at a fixed point in the sky instead. The words geostationary and geosynchronous are not the same geosynchronous requires just the orbit period, whereas geostationary also requires the orientation and shape of the orbit (Hasler, 1981). In several parts of the Clarke belt, nearby satellites apply a similar frequency band, and inside 2° of each other are located. A satellite aimed for radio networks among fixed earth stations should stay at a fixed point in the sky. It shows that the satellite should travel in a GEO (Figure 5.1).

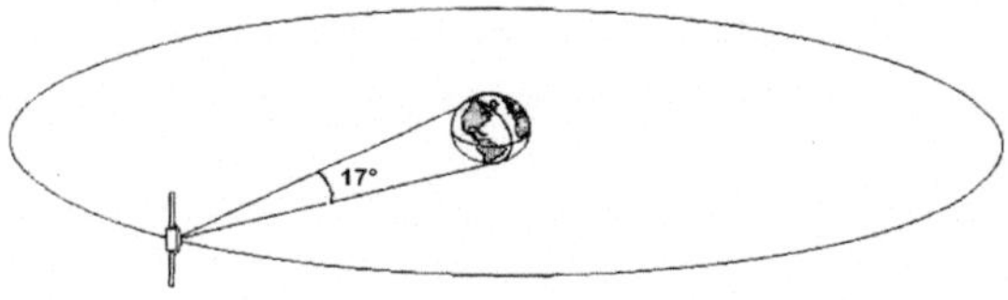

Figure 5.1. *The delimited angle of the Earth as understood from a satellite in the Clarke belt is approximately 17°.*

In two ways, the drift damages the satellite performance: the satellite may transfer outside position, or it may undertake an attitude, drift outcomes into external forces whereas there are several external forces substitute on the satellite, the prime forces are those employed by the sun and other objects of the solar system.

The direction and intensity of the gravitational field experienced by the sun constantly varies in 55-year cycles, yearly and daily. Its effect is being canceled by the cyclic nature of this force, at one pan of the cycle, an easterly pull is counterbalanced through a westerly pull 1/2 a day later: also, by a southerly pull, a northerly pull is counterbalancing nevertheless, there is a net resultant force which, over numerous months origins the satellite to move away from its geostationary position (Babuscia et al., 2013; Su *et al.*, 2017).

In the solar system, the pull of gravity of other objects is significantly lower than the sun's gravity. On Earth's surface, the rough spreading of landmass also roots mainly east-west drift to counter these forces, with some mechanism the satellite must be tailored to keep the satellite back into positions when it drifts (Lee *et al.*, 2016).

Two effects evident if the satellite is permitted to drift easily. First, about the Earth's equatorial plane, the orbit plane befits inclined. Over the Earth's center of gravity, the plane of orbit must cross, which shows that the satellite must travel over the Earth's equatorial plane double each sidereal day. For half of every sidereal day, the satellite is the north of the equatorial plane and for the other half south of it. Furthermore, due to the accumulation of angular momentum, the orbit believes an elliptic shape. Due to which the satellite no longer rotates at a constant velocity (Matricciani and Babuscia, 2012; Radhakrishnan et al., 2016). With small rockets, communication satellites are attached, known as thrusters, and a thruster is launched on command from a control station. Through its firing, it ejects a gas propulsive. All the parameters complicated in a firing is being measured by a ground station: the position of every thruster proportionate to the satellite, the duration, and timing of every pressure and fire of the ejected propellant; if these factors are measured properly, for years, the satellite can be upheld at the proper attitude and position this process is known as station-keeping (Walker, 1977; Matricciani and Riera, 2016).

Whenever a thruster is fired, the propellant is utilized once the amount of propellant is expended, the satellite cannot be retained at proper attitude and position, and the satellite must be discharged. Several types of research have been taken to describe the ideal tradeoff between propellant usage and satellite stability. These studies have demonstrated that a significant advantage of the propellant is applied for only one station keeping act maintaining the satellite from drifting near its north-south axis. The satellite drops the capability of station keeping when the propellant is done and becomes worthless to the satellite operator (Li and Liu, 2002; Matricciani, 2016).

In a GEO, since 1963, nearly 400 satellites have been positioned. Conventionally presuming an average time of 8 years per satellite, about 3200 years of in-orbit operation these satellites have collected. Already satellites are attaining an estimated operating service life of 15 years. This raises the probability that the satellite maintenance area will change (Katona et al., 2016; Vasavada *et al.*, 2016).

5.3. SATELLITE CONSTELLATIONS: MANY ARCHITECTURES AND MANY PROBLEMS

As per altitude, the orbits of satellite patterns can be split into LEO (low earth orbit), MEO (medium earth orbit), and GEO (geostationary Earth orbit) constellations. At any altitude, a Walker star constellation with polar orbits will be planned to apply to imitate the GEO with zenith paths at any latitude. If we consider the site at the equator, any receive/transmitter will be connected to a satellite and the satellite at the local zenith. Since the connected satellite is all the time viewed nearly at the local zenith, doppler phenomena are mostly minimized. The links between satellites are very easy since the positions in the orbital plane and nearby planes are constant, while with varying distances. No guiding antennas are necessary. The tropospheric proliferation fading and flashings are reduced (Fenech et al., 2016; Hasan and Bianchi, 2016).

Along with the altitude directly above the Earth surface, orbits of the satellite can be split up into LEO, MEO, and geostationary Earth orbits (GEO) (Qu et al., 2017; Su et al., 2019). In the equatorial plane (0° latitudes), a geostationary Earth orbits satellite is installed at around 36,000 km above the Earth's equator and it seems fixed to a fixed observer on the Earth. Due to the benefits of the fixed position, such as it does not need steering arrays or steering antennas, large coverage, and constant propagation lag, geostationary Earth orbits satellites are mainly applied to execute broadcasting and communication systems. Though at present, there is huge interest in giving high data-rate gain access to for the Internet services by use of geostationary Earth orbits satellites, and numerous geostationary Earth orbits high-output satellite communication systems have been positioned, for example, Inmarsat Global Xpress, Chinasat-16, Viasat IPStar, etc. Though, geostationary earth orbits communication systems suffer numerous disadvantages (Chotikapong et al., 2001; Di et al., 2019):

- For the present application, as nearly all Internet services need, the huge latency might not be endured;
- The free-space reduction is huge, thus, for mobile communications, it implies that the power needed for transmission ends in enhanced weight and size of consumer terminals, which raises the difficulty and lessens the serviceability of mobile terminals;
- At high latitudes, the service is not accessible since the path altitude angle is very low—the satellite can still be under the horizon—producing large tropospheric reduction because of raindrops, oxygen, and water vapor, mainly in the form of scintillation and rain;
- The geostationary Earth orbits orbit is exceptional, hence of restricted frequency and orbital resources-due to the smallest angle separation among nearby satellites in the geostationary Earth orbits orbit operating in the similar frequency band-that restrict the system capacity and coverage (Hasler, 1981);
- The charges of satellite launching and promoting it to the GEO are lofty, even though it can be reduced via launching after equatorial sites. Nevertheless, the total cost of positioning and launching satellites in the GEO for worldwide coverage may be beneath that of positioning and launching hundreds or even more of low earth orbit satellites (LEOS). Furthermore, additional charges are related to the ground control of satellites, decrease in geostationary Earth orbits patterns, greater in LEO constellations.

REFERENCES

1. Babuscia, A., Corbin, B., Jensen-Clem, R., Knapp, M., Sergeev, I., Van, D. L. M., & Seager, S., (2013). Commcube 1 and 2: A CubeSat series of missions to enhance communication capabilities for CubeSats. In: *Proceedings of the 2013 IEEE Aerospace Conference, Big Sky* (Vol. 2013, pp. 1–19). MT, USA.
2. Chotikapong, Y., Cruickshank, H., & Sun, Z., (2001). Evaluation of TCP and internet traffic via low earth orbit satellites. *IEEE Pers. Commun., 8*, 28–34. [CrossRef.].
3. Clarke, A. C., (1966). Extra-terrestrial relays: Can rocket stations give worldwide radio coverage? In: *Progress in Astronautics and Rocketry* (Vol. 19, pp. 3–6). Elsevier.
4. Del, P. I., Cameron, B. G., & Crawley, E. F., (2019). A technical comparison of three low earth orbit satellite constellation systems to provide global broadband. *Acta Astronautica, 159*, 123–135.
5. Di, B., Song, L., Li, Y., & Poor, V. H., (2019). Ultra-dense LEO: Integration of satellite access networks into 5G and beyond. *IEEE Wirel. Commun., 26*, 62–69. [CrossRef.].
6. Draim, J. E., Cefola, P., & Castrel, D., (2000). Elliptical orbit constellations-a new paradigm for higher efficiency in space systems? In: *2000 IEEE Aerospace Conference. Proceedings (Cat. No. 00TH8484)* (Vol. 7, pp. 27–35). IEEE.
7. Fenech, H., Tomatis, A., Amos, S., Soumpholphakdy, V., & Serrano, M. J. L., (2016). Eutelsat HTS systems. *Int. J. Satell. Commun. Netw., 34*, 503–521. [CrossRef.].
8. Hasan, M., & Bianchi, C., (2016). Ka-band enabling technologies for high throughput satellite (HTS) communications. *Int. J. Satell. Commun. Netw., 34*, 483–501. [CrossRef.].
9. Hasler, A. F., (1981). Stereographic observations from geosynchronous satellites: An important new tool for the atmospheric sciences. *Bulletin of the American Meteorological Society, 62*(2), 194–212.
10. Ineichen, P., & Perez, R., (1999). Derivation of cloud index from geostationary satellites and application to the production of solar irradiance and daylight illuminance data. *Theoretical and Applied Climatology, 64*(1), 119–130.

11. Johnston, E., (2013). *List of Satellites in Geostationary Orbit* (Vol. 1, pp. 1–44). http://www.satsig.net/sslist.htm (accessed on 20 April 2021).
12. Katona, Z., Clazzer, F., Shortt, K., Watts, S., Lexow, H. P., & Winduratna, R., (2016). Performance, cost analysis, and ground segment design of ultra-high-throughput multi-spot beam satellite networks applying different capacity enhancing techniques. *Int. J. Satell. Commun. Netw., 34*, 547–573. [CrossRef.].
13. Knepp, T., Pippin, M., Crawford, J., Chen, G., Szykman, J., Long, R., & Neil, D., (2015). Estimating surface NO_2 and SO_2 mixing ratios from fast-response total column observations and potential application to geostationary missions. *Journal of Atmospheric Chemistry, 72*(3), 261–286.
14. Kodheli, O., Guidotti, A., & Vanelli-Coralli, A., (2017). Integration of satellites in 5G through LEO constellations. In: *GLOBECOM 2017–2017 IEEE Global Communications Conference* (Vol. 1, pp. 1–6). IEEE.
15. Lee, S., Wu, Y., & Mortari, D., (2016). Satellite constellation design for telecommunication in Antarctica. *Int. J. Satell. Commun. Netw., 34*, 725–737. [CrossRef.].
16. Leese, J. A., Novak, C. S., & Clark, B. B., (1971). An automated technique for obtaining cloud motion from geosynchronous satellite data using cross-correlation. *Journal of Applied Meteorology and Climatology, 10*(1), 118–132.
17. Li, S. Y., & Liu, C. H., (2002). An analytical model to predict the probability density function of elevation angles for LEO satellite systems. *IEEE Commun. Lett., 6*, 138–140.
18. Liu, G. R., & Lin, T. H., (2004). Application of geostationary satellite observations for monitoring dust storms of Asia. *Terrestrial, Atmospheric and Oceanic Sciences, 15*(5), 825–837.
19. Malik, A. Q., Mufti, A., Hiser, H. W., Veziroglu, N. T., & Kazi, L., (1991). Application of geostationary satellite data for determining solar radiations over Pakistan. *Renewable Energy, 1*(3, 4), 455–461.
20. Matricciani, E., & Babuscia, A., (2012). A belt of satellites in low equatorial orbits simulating the geostationary orbit. In: *Proceedings of the 30th AIAA International Communications Satellite Systems Conference (ICSSC)* (Vol. 30, pp. 1–15). Ottawa, ON, Canada.

21. Matricciani, E., & Riera, J. M., (2016). Variable elevation-angle radio links faded by rain at K_a-Band from Madrid to the sun-earth Lagrangian point L1. In: *Proceedings of the 22th Ka and Broadband Communications Conference* (Vol. 1, pp. 1–33). Cleveland, OH.
22. Matricciani, E., (2016). Space communications with variable elevation angle faded by rain: Radio links to the sun-earth first Lagrangian point L1. *Int. J. Satell. Commun. Netw., 34*, 809–831. [CrossRef.].
23. Picone, J. M., Emmert, J. T., & Lean, J. L., (2005). Thermospheric densities derived from spacecraft orbits: Accurate processing of two-line element sets. *Journal of Geophysical Research: Space Physics, 110*(A3), 1–33.
24. Qu, Z., Zhang, G., Cao, H., & Xie, J., (2017). LEO satellite constellation for internet of things. *IEEE Access, 5*, 18391–18401. [CrossRef.].
25. Radhakrishnan, R., Edmonson, W. W., Afghah, F., Rodriguez-Osorio, R. M., Pinto, F., & Burleigh, S. C., (2016). Survey of inter-satellite communication for small satellite systems: Physical layer to network layer view. *IEEE Commun. Surv. Tutor., 18*, 2442–2473. [CrossRef.].
26. Rao, P. K., Holmes, S. J., Anderson, R. K., Winston, J. S., & Lehr, P. E., (1990). *Weather Satellites: Systems, Data, and Environmental Applications, 1,* 1–33.
27. Sánchez, A. H., Soares, T., & Wolahan, A., (2017). Reliability aspects of mega-constellations satellites and their impact on the space debris environment. In: *2017 Annual Reliability and Maintainability Symposium (RAMS)* (Vol. 1, pp. 1–5). IEEE.
28. Su, W., Lin, J., & Ha, T., (2017). Global communication coverage using CubeSats. In: *Proceedings of the IEEE 7th Annual Computing and Communication Workshop and Conference (CCWC)* (Vol. 2017, pp. 1–7). Las Vegas, NV.
29. Su, Y., Liu, Y., Zhou, Y., Yuan, J., Cao, H., & Shi, J., (2019). Broadband LEO satellite communications: Architectures and key technologies. *IEEE Wirel. Commun., 26*, 55–61. [CrossRef.].
30. Vasavada, Y., Gopal, R., Ravishankar, C., Zakaria, G., & BenAmmar, N., (2016). Architectures for next generation high throughput satellite systems. *Int. J. Satell. Commun. Netw., 34*, 523–546. [CrossRef.].
31. Walker, J. G., (1977). *Continuous Whole-Earth Coverage by Circular-Orbit Satellite Patterns* (pp. 1–30). Royal Aircraft Establishment: Farnborough, UK, Technical Report 77044.

32. Wang, J., Xu, X., Ding, S., Zeng, J., Spurr, R., Liu, X., & Mishchenko, M., (2014). A numerical testbed for remote sensing of aerosols, and its demonstration for evaluating retrieval synergy from a geostationary satellite constellation of GEO-CAPE and GOES-R. *Journal of Quantitative Spectroscopy and Radiative Transfer, 146*, 510–528.
33. Yang, H., Oh, E., Choi, J. K., Park, Y. J., & Han, H. J., (2016). Application of geostationary ocean color imager data to the extraction of ocean fronts. *Remote Sensing Letters, 7*(5), 456–465.
34. Zhang, X., Kondragunta, S., Ram, J., Schmidt, C., & Huang, H. C., (2012). Near-real-time global biomass burning emissions product from geostationary satellite constellation. *Journal of Geophysical Research: Atmospheres, 117*(D14), 1–30.
35. Zhu, H., Zhao, H. L., Jiang, Y. Z., & He, T., (2013). Overview of the application of geostationary satellites in water conservancy. *South-to-North Water Transfers and Water Science and Technology, 4*, 1–33.

Chapter

6

AN OVERVIEW OF CUBE-SATELLITE PROPULSION TECHNOLOGIES AND TRENDS

CONTENT

6.1. Introduction.......126

6.2. Cold Gas Propulsion (CGP) Systems.......128

6.3. Liquid Propulsion (LP) Systems.......131

6.4. Solid Rocket Propulsion (SRP) Systems.......135

6.5. Resistojets.......138

6.6. Radio-Frequency Ion Thruster (RIT).......141

6.7. Hall Effect Propulsion/Hall Thrusters.......144

References.......147

6.1. INTRODUCTION

Current years have observed an incredible rise in curiosity in the use of Cube-Satellites (CubeSats) among the space community comprising space agencies, the industry as well as academia. Two factors have swayed this squirt of interest: first, cost-effective entree to space as a secondary payload for technology validation, science proof-of-concept substantiation, communication, and education; second, utilizing commercial-off-the-shelf (COTS) technologies in the design architecture. These two aspects have directed to a substantial low inclusive cost of a CubeSat mission (Woellert et al., 2011). Several commercial companies propose services to launch typical (1U–3U) CubeSats, and they charge between $50,000 to $200,000 per CubeSat based on its design size and altitude of deployment (Nervold et al., 2016). Lockheed Martin, SpaceX, Boeing, Interorbital Systems, and Virgin Galactic have declared to minimize this expense to somewhere in the range of $10,000 and $85,000 by 2020. Numerous space agencies also offer CubeSat launch opportunities to academic circles through different proposal solicitations, ensuing in Universities across the globe launching their CubeSats and delivering appreciated space systems engineering training to students (Barnhart *et al.,* 2007). CubeSats have principally flown in 1U and 3U form factors; however, CubeSats beyond 3U are not unusual; other platforms that have been deliberated comprise 6U (12 kg), 12U (24 kg), and 27U (54 kg) (Toorian et al., 2005; Hevner *et al.,* 2011).

Up to now, CubeSats have been exploited solitary for near-earth missions, though, a few far Earth and interplanetary missions (INSPIRE and MarCO) have also been anticipated]. CubeSats have primarily been constrained in their procedures due to their small size that confines their on-board competencies (power, mobility, and payload) leading to inadequate mission life and range of travel (after deployment). There are various sub-systems onboard a CubeSat, comprising payload, communication, and data handling, mobility (propulsion and attitude control systems), and power systems (Klesh *et al.*, 2013). W-5 A propulsion system is the primarily mobility system of a spacecraft and supports various maneuvering operations like orbit changing and station keeping. A crucial parameter that discriminates against a propulsion system is its reliance on onboard power. Consequently, propulsion systems can be categorized into two forms: electric and non-electric systems. Electric propulsion systems are generally categorized into resistojet, electrospray, ion, Hall, and pulsed plasma systems, and they dynamically entail on-board power for their operation, while the non-electric propulsion systems can be categorized into cold gas, liquid, and solid rocket systems, and they involve on-board power only to regulate (initiate and terminate) the propulsion process (Schoolcraft et al., 2017).

Afterward, four key performance factors for any propulsion system are defined: Specific impulse (I_{sp}), thrust (τ), effective exit velocity (or exit velocity) (v_e), and delta-v (Δv). It is important to understand these elements to well realize the operation of a propulsion system. The thrust (shown in Eqn. (1)) produced is a combination of momentum thrust and pressure thrust. Momentum thrust is influenced by the mass flow rate ($\dot{m}$) of propellant and the exit (exhaust) velocity (v_e) whereas, pressure thrust, instead, is a function of the exit area (A_e), exit pressure (P_e), and ambient pressure (P_a) (Sutton and Biblarz, 2001; Kolmas et al., 2016). The ambient pressure for the case of spacecraft propulsion systems is estimated to zero because of the vacuum environments experienced in space. Nozzles form the expansion zones for the propellants (in cold gas, liquid, solid rocket, and resistojet systems) and their geometry shows a substantial role in accelerating the propellants and in creating high thrust (higher than typical electric propulsion systems). A converging-diverging (CD) type nozzle or *de Laval* nozzle is extensively used since it transforms a larger fraction of the energy existing in the propellants into kinetic energy. The gases passing over a CD nozzle can break the sound barrier (Mach number > 1), and henceforth, they are also denoted as supersonic nozzles. Specific impulse (shown in Eqn. (2)) is the impulse (integral of thrust over time) produced per unit weight (at sea level) of propellant and is reliant on the thrust created and mass flow rate of the propellant ($\dot{m}$). Exit velocity (shown in Eqn. (3)) is the velocity of the propellant at the exit area of the nozzle and can be deliberate from the product of the specific impulse and acceleration due to gravity at sea level (g_o). The delta-v (shown in Eqn. (4)) is acquired from the well-known *Tsiolkovsky Rocket Equation* that relates the exit velocity of a spacecraft to its initial (m_i) and final (m_f) masses. The relationships between the performance factors are abridged below (Bowen et al., 2015; Roscoe et al., 2015):

$$\tau = \dot{m}v_e + (P_e - P_a)A_e \tag{1}$$

$$I_{sp} = \frac{\tau}{\dot{m}g_o} \tag{2}$$

$$v_e = g_o I_{sp} \tag{3}$$

$$\Delta v = v_e ln\left(\frac{m_i}{m_f}\right) \tag{4}$$

In this section, micro-propulsion systems have been investigated, some of which have flown on CubeSats, whereas others are in the development phase and are in prospective concern for future CubeSat missions. A subset of these has flown on bigger satellites as secondary propulsion systems. Based on the survey conducted, the performance factors of micro-propulsion systems have been briefed in terms of: first, a comparison of thrust and specific impulse for all propulsion systems; second, a comparison of power and specific impulse, as also thrust-to-power ratio, and specific impulse for electric propulsion systems. Some studies also exist in the literature summarizing the high-tech micro-propulsion technologies: a contemporary survey on micro-propulsion systems—Lemmer (2017) and NASA Mission Design Division Report—Small Spacecraft Technology State of the Art (Mission Design Division, 2015).

6.2. COLD GAS PROPULSION (CGP) SYSTEMS

6.2.1. Operating Principle

A cold gas propulsion (CGP) System depends on the process of controlled ejection of compressed liquid or gaseous propellants to produce thrust. Because of the nonexistence of a combustion process, a CGP system entails only one propellant (without an oxidizer) and therefore can be deliberate with minimum complexity. The representation of a typical CGP system is presented in Figure 6.1, and the key components comprise propellant storage and a nozzle.

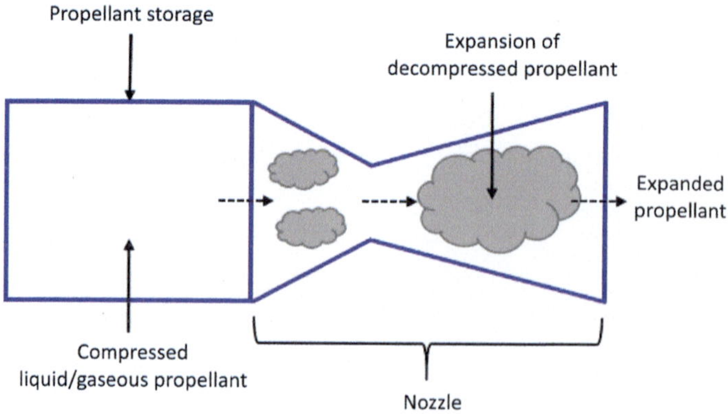

Figure 6.1. *Schematic of a cold gas propulsion system.*

The simpler design of a CGP system points to a smaller system mass and lower power necessities for regulation purposes. Though, these benefits come at the expense of a monotonically declining thrust profile over a certain period. The thrust created is directly proportional to the pressure of the propellant inside the tank (propellant storage) and throughout the mission, tank pressure decreases (due to propellant usage) causing a decline in the maximum thrust that is generated by the system (Persson *et al.*, 2005; Wu *et al.*, 2014).

Specific impulse (shown in Eqn. (5)) of a CGP system is primarily influenced by the exit-to-chamber-pressure (P_e/P_c) and characteristic velocity (C^*). The exit-to-chamber-pressure is associated with the expansion of the propellant, while Poisson constant (γ) is the ratio of specific heat at constant pressure and constant volume. The characteristic velocity of a CGP system at any instant of time is a function of the velocity of propellant in Mach number (Anis, 2012). Exit velocity (as shown in Eqn. (6)) is an additional main performance factor that not only depends on the exit-to-chamber-pressure but also the chamber temperature (T_c). The mathematical relations are described below (Bonin et al., 2015; Manzoni and Brama, 2015):

$$I_{sp} = \frac{\gamma C^*}{g_0}\sqrt{\frac{2}{\gamma-1}\left(\frac{2}{\gamma+1}\right)^{\frac{\gamma+1}{\gamma-1}}\left(1-\frac{P_e}{P_c}\right)^{\frac{\gamma-1}{\gamma}}} \qquad (5)$$

$$v_e = \sqrt{\frac{2\gamma T_c}{\gamma-1}\left(1-\frac{P_e}{P_c}\right)^{\frac{\gamma-1}{2}}} \qquad (6)$$

6.2.2. Design Considerations and Technologies

A CGP system can use either liquid or gaseous propellants, though, using a liquid propellant will result in a drop in the storage volume. The propellant selected must have high-density-I_{sp} (Specific impulse/unit volume) to rise the longevity of the onboard propellant. Moreover, the lesser storage pressure of the propellant assists the design of storage tanks with greater safety boundaries. The harmfulness and the easiness of obtainability of the propellant also influence the design cost of a propulsion system during the on-ground development and assembly operations of the spacecraft. So, using eco-friendly propellants will bring about a decrease in expenses incurred for safety procedures, storage, and transportation (Gibbon, 2010). Liquid propellants deliver the benefit of lessening the storage volume, though they can give rise to a de-stabilizing effect because of the sloshing of propellant inside the tank. Although no explicit sloshing deterrence technology is existing for CubeSats, special anti-sloshing baffles technology has been utilized in a micro-propulsion system of a bigger satellite; these baffles are employed to detain the flow of propellant and have been exploited in the SNAP-1 propulsion system produced by Surrey Satellite Technology Ltd. (SSTL), Guildford, UK for the Giove-A mission (Bauer, 1963).

A currently studied model called solar thermal propulsion system has the potential of employing solar energy in refining the performance of CGP systems. In this system, concentrated solar energy is utilized to directly heat the propellant. As the propellant moves in the nozzle at an elevated temperature, a considerably enhanced thrust and an amplified specific impulse comparative to a regular cold gas flow are witnessed. The solar thermal propulsion system concept was certified for a larger spacecraft, for example, the orbital station-keeping scenario for a 200 kg spacecraft in a circular orbit was deliberated.

Table 6.1. Summary of Cold Gas Propulsion Systems

Company/ Institution with Location	Engine	Propellant	I_{sp} (s)	Thrust (mN)	Heritage	Remarks
UTIAS-SFL, Toronto, ON, Canada	CNAPS	SF_6	<35	10–40	CanX-4 (6 kg), CanX-5 (6 kg)	–
GOMSpace, Denmark	MEMS Cold Gas	Methane	50–75	1	TW-1 (one 3U and two 2U)	Also flown on PRISMA (180 kg)
VACCO Industries, El Monte, CA, USA	CPOD	R134a	40	25	CPOD (3U)	–
SSTL, Guildford, UK	SNAP 1	Liq. Butane	43	50	–	Flown on Giove-A (600 kg)
Microspace Rapid, Singapore	POPSAT-HIP1	Argon	43	1	POPSAT-HIP1 (3U/3.3 kg)	–

The orbital parameters for this spacecraft were 600 km altitude, 28.5° inclination, and 1 km decay/day, and, it required a one-minute burn of 1.9 N thrust to counter the orbital decay (Reid et al., 2013). The 1.9 N burn was attained with a specific impulse of 300 s, 0.64 g/s mass flow rate of propellant, and an exit flow temperature of 1500 K. A foremost problem of this technology is its reliance on direct solar illumination at the time of propulsive maneuvers. Table 6.1 offers a summary of the performance parameters of the surveyed cold gas systems. As it is obvious, CGP systems have been extensively utilized on CubeSat missions; only one of the systems does not have CubeSat heritage but has flown on a larger satellite (Underwood *et al*., 2003; Bradford *et al*., 2006).

6.3. LIQUID PROPULSION (LP) SYSTEMS

6.3.1. Operating Principle

In a liquid propulsion (LP) system, thrust is produced by way of expelling the gases formed through the process of combustion of liquid propellant(s). Based on the mission requirements, a spacecraft can have LP systems with one (mono) or two (bi) propellants.

Mono-propellant LP systems utilize a catalyst to decompose (burn) the propellant and create thrust. The decomposition process occurs when the propellant is injected into the combustion chamber through the catalyst bed. Illustrations of mono-propellants are hydrazine and nitrous oxide] and that of a catalyst are solid manganese dioxide, liquid permanganates, platinum, and iron oxide.

A bi-propellant LP system, alternatively, includes both oxidizer and fuel. Combination of liquid oxygen and kerosene, or liquid oxygen and RP1 are examples of bi-propellants that are extensively used (Ley et al., 2009). The fuel in the bi-propellant system can occasionally be used in a mono-propellant context with the addition of a catalyst. Either LP systems have principally been used on larger satellites for high Δv (orbit-raising) operations and a single propellant is normally used for low Δv operations (station keeping). The representation of a bi-propellant LP system is presented in Figure 6.2 and it principally comprises a combustion chamber, nozzle, and propellant storage for both oxidizer and fuel. In the case of a mono-propellant system, the key components are propellant storage (only fuel), catalyst bed, and the nozzle.

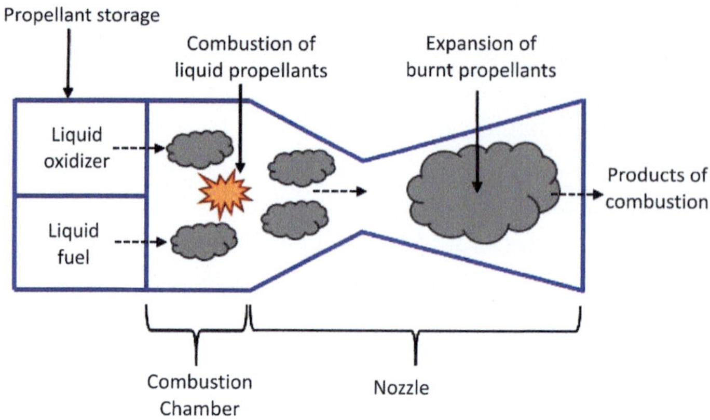

Figure 6.2. *Schematic of a bi-propellant liquid propulsion system.*

The thrust and specific impulse of an LP system can be achieved from Eqns. (1) and (2) respectively. Exit velocity (shown in Eqn. (7)) of an LP system, like a CGP system, is reliant on the exit-to-chamber-pressure-ratio (P_e/P_c) and combustion chamber temperature (T_c).

$$v_e = \sqrt{\frac{2\gamma}{\gamma-1}\Re T_c\left[1-\left(\frac{P_e}{P_c}\right)^{\frac{\gamma-1}{\gamma}}\right]}, \quad (7)$$

where; γ is the Poisson constant; and R is the universal gas constant.

6.3.2. Design Considerations and Technologies

LP systems, like CGP systems, deal with problems associated with the storage and operational pressures of the propellant. To this end, the maximum expected operating pressure (MEOP), which is the maximum possible pressure at which the propellant is anticipated to operate, is a significant design parameter. MEOP of propellant has to be enviably high so that thruster performance (thrust, specific impulse) can be increased (Stratton, 2004). Extremely toxic propellants like hydrazine have been effectively utilized in larger spacecraft for over 60 years. Currently, there has been a worldwide emphasis on the development and usage of lower-toxic green propellants to lessen hazards experienced because of contamination during laboratory testing and mission phases while in space. Green mono-propellants (GEM) are less harmful because of one of two reasons: either as a result of their non-threatening toxicology even for probable levels of unintentional ingestion, or their low vapor pressure posturing no substantial risk of being inhaled. Several emergent green propellants (ammonium dinitramide (ADN), Sulfur Hexafluoride (SF_6), AF-M315E) deliver substantial supplementary advantages like better physical characteristics (higher density), improved performance for the propulsion system (higher thrust and specific impulse), and abridged thermal conditioning requirements for storage compared to hydrazine. Yet, they do show a disadvantage with requiring greater preheat temperatures, higher than the typical 120–150°C of hydrazine thrusters (Sackheim and Masse, 2014).

Newly developed AF-M315E is a high-performance Hydroxyl-Ammonium Nitrate (HAN) based green propellant (optimum mixture stability despite being a low-toxicity hazard) developed by US Air Force Research Laboratory (AFRL), Wright-Patterson Air Force Base, OH, USA. AF-M315E propellant contributes 50% higher density-specific-impulse (specific impulse/unit volume) compared to the hydrazine and is yet being tested for CubeSat applications; nevertheless, it is yet to fly on any spacecraft (Spores, 2015).

The least storage and operating temperatures of AF-M315E possibly mark it a high-interest application in cold environments where unremitting thermal conditioning might not be feasible.

Moreover, a system utilizing AF-M315E propellant will not probably encounter design concerns and failure modes related to control of mixture ratio of propellant, vapor diffusion and reaction, and oxidizer flow decay. A foremost disadvantage of AF-M315E is that it is hard to ignite because of its ionic liquid (IL) (high water content) form (Whitmore, 2017). Multiple catalyst systems were used to conduct experiments to enhance its ignitability, but room temperature ignition does not presently exist, and the preheating process can devour a large amount of energy (up to 15,000 J). These requirements on energy utilization execute unadorned limitations on spacecraft and even more so on CubeSats.

To accommodate the requirements of diverse CubeSat missions and to escalate their lifespan, micro-propulsion system developers have cropped up with form-factor customization based on the quantity of on-board propellants that can be carried. MPS-120 CHAMPS, BGT-X5, HPGP, and VACCO/ECAPS are examples of micro-propulsion systems designed in numerous configurations varying from 0.5 U to 2 U. For a specified system, the difference in configurations consequences typically in the quantity of propellant they carry. However, none of the above have flown on a CubeSat;

though, HPGP micro-propulsion system has flown on two larger spacecraft. Table 6.2 delivers a summary of the surveyed LP systems and their performance factors. Note that, though HYDROS (developed by Tethers Unlimited, Bothell, WA, USA) is a hybrid electric/chemical propulsion system, yet because the propellant is water, it is summarized in Table 6.2 (Friedhoff et al., 2017; Tsay et al., 2016, 2017).

Table 6.2. Summary of Liquid Propulsion Systems

Company/Institution with Location	Engine	I_{sp} (s)	Thrust (mN)	Propellant	Remarks
ECAPS, Solna, Sweden	HPGP	231–232	1000	ADN based LMP-103S	Flown on PRISMA (180 kg) and SkySat-3 (10.5 kg tank)
Aerojet Rocketdyne, Sacramento, CA, USA	MPS-120 CHAMPS	215	260	Hydrazine	–
Busek, Natick, MA, USA	BGT-X1	214	100	AF-M315E	–
Aerojet Rocketdyne, Sacramento, CA, USA	MPS-130 CHAMPS	240	1.5	AF-M315E	–
Tethers Unlimited, Bothell, WA, USA	HYDROS	256	250–600	Liquid water	–
Aerojet Rocketdyne, Sacramento, CA, USA	GPIM propulsion system	235	400–1100	AF-M315E	–
Busek, Natick, MA, USA	BGT-X5	220–225	500	AF-M315E	–

6.4. SOLID ROCKET PROPULSION (SRP) SYSTEMS

6.4.1. Operating Principle

A solid rocket propulsion (SRP) system operates on the principle of burning solid propellants and producing thrusts by expelling the gases formed during combustion. Like LP bi-propellant system, an oxidizer is utilized in the SRP system. Nevertheless, it varies from an LP system in different ways: firstly, the solid propellants are stored within the combustion chamber itself; secondly, sloshing effects observed in LP systems are absent since both fuel and oxidizer are solids. Though SRP systems do not experience sloshing, the deficiency of control over propellant burn rate generates difficulty for thrust regulation. The schematic of an SRP system is displayed in Figure 6.3 and consists of a combustion chamber that holds the solid propellant, an igniter that starts the combustion process, and a nozzle (Schmuland et al., 2011; Kolosa et al., 2014).

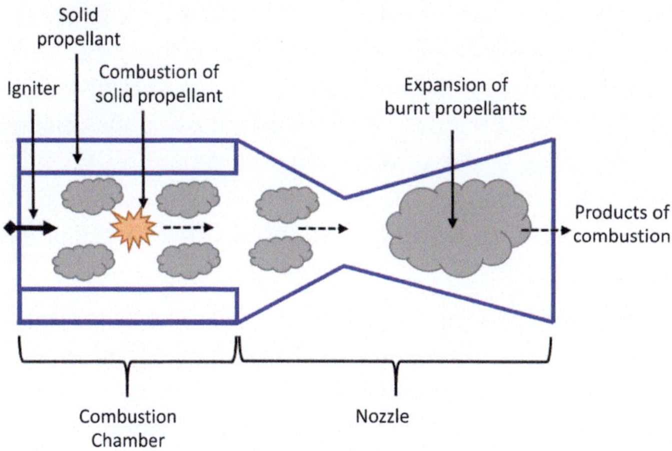

Figure 6.3. *Schematic of a solid propulsion system.*

Since thrust regulation is challenging in SRP systems, burn rate can be utilized in the initial phase of system design to apprehend the combustion process as it manages the mass flow rate of hot gases produced during combustion. The burn rate (r) (shown in Eqn. (8)) is reliant on the chamber pressure (P_c), temperature coefficient (a), and combustion index (n). The temperature coefficient is a non-dimensional empirical constant, whereas the combustion index defines the effect of chamber pressure on the burn rate. For a propulsion system equipped with a *de Laval* (CD) nozzle, the characteristic velocity (C^*) (shown in Eqn. (9)) relates to the productivity of the combustion process and is independent of nozzle characteristics. The thrust, specific impulse, and exit velocity of an SRP system can be deliberated the same way as done for LP systems from Eqns. (1), (2), and (7), respectively. The mathematical relations summarizing burn rate and characteristic velocity are defined below (Carpenter et al., 2013; Spores et al., 2013):

$$r = aP_c^n \tag{8}$$

$$C^* = \frac{P_c A_t}{\dot{m}} \tag{9}$$

6.4.2. Design Considerations and Technologies

The problems with thrust regulation in SRP systems, a distinctive addition (to the existing SRP system design) was suggested and designed by Aerospace Corporation, El Segundo, CA, the USA for the I_{sp} 30 s Motor SRP system. This addition comprises an external movable mass (pitch/yaw system) with 8 jet paddles. The jet paddles are placed just after the nozzle and constitute rectangular moving arms (plates or slabs) with one of their faces exposed to the exhaust flow. Thrust regulation arises by adjusting the orientation of the paddles and imparting preferred directionality to the flow (Zondervan et al., 2014). Further, this technology could significantly initiate benefits for other micro-propulsion systems as well, explicitly LPS (Schmuland *et al.*, 2012; Legge *et al.*, 2017).

A drawback of SRP systems is their one-shot use because of deficiency of control over propellant burn rate. To alleviate this disadvantage, a system of hundreds of Solid Propellant Micro-thrusters (SPMs) has been suggested by Sathiyanathan et al. (2011); these micro-thrusters can be utilized by forming a tightly spaced matrix (within the constraints of available external surface area). In SPMs, the solid energetic propellant is burnt (during the combustion process) and the resultant gases are accelerated through micro-nozzles. The size of the thruster can be amended to suit the thrust requirements, and programmable thrust delivery can be attained through the instantaneous or successive firing of multiple thrusters. A typical SRP micro-thruster makes use of MEMS technology and includes numerous laminated layers comprising a combustion chamber, a nozzle, an igniter, and a seal. The combustion chamber stores the solid energetic-propellant and the igniter section heats the propellant using a resistive heating element.

Silicon or nichrome are commonly used as materials for the heating element (Rossi et al., 2006). Nozzles are designed to encounter mission-specific thrust requirements and *de-Laval* (CD) nozzles are usually selected for their higher performance.

The seal comprises an epoxy or similar material or mechanisms. Likewise, the silicon wafer is used in these micro-thrusters since it increases the ignition efficiency by reducing the current leakage (Zhang et al., 2004). Besides addressing the concern of minimal control over the propellant burn, SPMs evade supplementary external-surface-area requirements in consequence of the use of a traditional nozzle (instead micro-nozzles are distributed over the surface of the spacecraft). Moreover, SPMs deliver the proficiency to produce differing torque values reliant on thruster distance from the center of mass (James *et al*., 2015).

An alternate technology that delivers an improvement of burn rate regulation was suggested by digital solid-state propulsion (DSSP), Reno, NV, USA, through the creation of a new electric solid propellant (ESP). ESP is a hydroxyl ammonium nitrate (HAN) based GEM and proposes higher theoretical performance compared to AF-M315E propellant (discussed earlier in LP systems section).

This innovative technology is integrally safe since ignition is possible only through an uninterrupted supply of electrical power, thus plummeting the chances of flames produced due to accidents, and it possesses the potential to be used as a propellant in both chemical and electrical propulsion systems (Thrasher *et al.,* 2016). This technology has flown on a larger satellite; SPINSAT, a spherical satellite with a 22-inch diameter and a mass of 57 kg, was launched in 2014 and housed 72 DSSP thrusters (Nicholas *et al*., 2013).

Several economical alternative propellants have been investigated for SRP systems and in recent times, aluminum wool as a propellant, together with the mixture of sodium hydroxide and water as an oxidizer, was tested and was found to yield a thrust of 32 mN and a specific impulse of 45 s. A foremost benefit of these propellants is that they are cost-effective, easy to handle, and can be stored over a prolonged duration deprived of any decomposition (David and Knoll, 2017). Table 6.3 delivers a summary of the surveyed SRP systems and their performance parameters.

All of these systems do not have a heritage of flying on a CubeSat mission, though, as already specified, one of these, the DSSP CAPS-3 propulsion system has flown on the SPINSAT mission (Mueller et al., 2008; Sawka and McPherson, 2013).

Table 6.3. Summary of Solid Propulsion Systems

Company/Institution with Location	Engine	I_{sp} (s)	Thrust (N)	Propellant	Remarks
DSSP, Reno, NV, USA	CDM-1	226	76	AP/HTPB	–
Aerospace Corporation, El Segundo, CA, USA	I_{sp} 30 s	187	37	–	–
DSSP, Reno, NV, USA	CAPS-3	245–260	–	HIPEP-501A	Flown on SPINSAT (57 kg)
Orbital ATK, Dulles, VA, USA	STAR 4G	269.4	13	Al and Ammonium perchlorate	–

6.5. RESISTOJETS

6.5.1. Operating Principle

In a resistojet, the propellant is being passed via a heat exchanger (or heating element) where it is super-heated and expelled through an expansion nozzle. For example, laboratory experiments have revealed exit temperatures of 600–1050°C for methanol and 300–1175°C for ammonia propellants (Frisbee, 2003). The heating process lessens the gas (propellant) flow rate from a given upstream pressure via a given nozzle area, hence leading to the escalation in the specific impulse that is proportionate to the square root of temperature as portrayed in Eqn. (11) (Martinez-Sanchez and Pollard, 1998; Robin et al., 2008).

The working principle of a resistojet is analogous to that of a CGP system excluding that the propellant is heated earlier than the expansion process. Despite the propellant's high energy (gained by heating), an exhaust velocity considerably greater than that of a CGP system is attained in a resistojet. Exit velocities of micro CGP systems range between around 300–700 m/s; whereas those of micro resistojets are almost 2.2 km/s (Slough et al., 2005; Chianese and Micci, 2006). A major disadvantage of resistojets is that their performance (thrust, I_{sp}) is restricted by the melting temperature of the heating element used. Additionally, power and thermal losses during heating of the elements add to the inadequacy of resistojets. The schematic of a resistojet propulsion is presented in Figure 6.4, and the leading components comprise propellant storage, heating element, and nozzle (Matticari et al., 2006; Skuhersky et al., 2017).

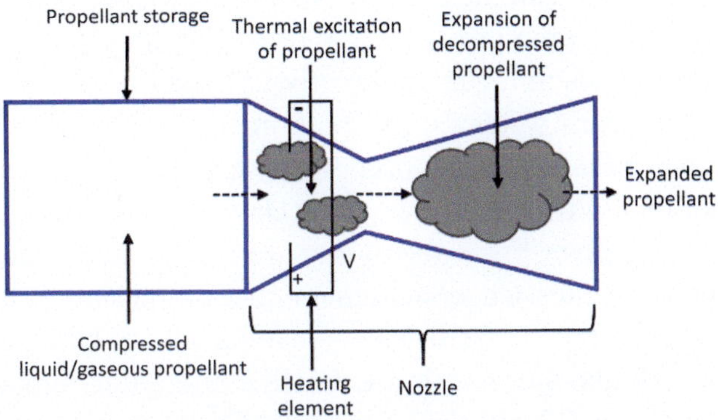

Figure 6.4. *Schematic of a Resistojet propulsion.*

The thrust (shown in Eqn. (10)) created by the propellant at stagnation pressure is also determined by stagnation number density of propellant (n_o) in m^{-3}, stagnation temperature (T_o), and the probability (χ) of a molecule exiting the expansion slot area (A_o). Specific impulse (shown in Eqn. (11)) is a function of the stagnation temperature and the mass of the propellant (m) (Ketsdever et al., 2001).

$$\tau = A_o\left(\frac{n_o k T_o}{2}\right)\chi \tag{10}$$

$$I_{sp} = \sqrt{\frac{\pi k T_o}{2m}}\frac{1}{g_o}, \tag{11}$$

where; k is the Boltzmann constant; and g_o is the acceleration due to gravity at sea level.

6.5.2. Design Considerations and Technologies

Resistojets with a range of propellants have been utilized on larger satellites, and similar to any other systems with liquid propellants, they have experienced problems because of sloshing within the tanks (Lee *et al*., 2008). Free molecule micro resistojet (FMMR) developed with water as the propellant is one of the systems that can resolve these problems. FMMR is cost-effective, low power consumption, and low mass MEMS fabricated resistojet that functions by heating a propellant gas as it expands through a series of slots.

There are three key advantages of using water as the propellant: firstly, water is stored as a liquid and to save the volume occupied by the propellant as a result of its high storage density; secondly, because of its lower molecular mass, water propellant can increase the specific impulse; thirdly, water has an adequately high vapor pressure at typical SmallSat (<10 kg) on-orbit temperature due to which it can be directly utilized to create thrust without pre-vaporization. During laboratory experiments of FMMR, it has been revealed that the influence of the propellant sloshing on spacecraft attitude stability is nominal (Ahmed *et al.*, 2006; Davies *et al.*, 2012). Resistojets are recognized to deliver lower thrust and are chiefly employed for attitude control on larger satellites. CubeSat high impulse propulsion system (CHIPS) resistojet developed by CU Aerospace, Champaign, IL, USA, and VACCO Industries Inc., Huntsville, AL, USA. CHIPS offers dual-mode operation: first, warm fire mode (30 mN thrust and 82 s I_{sp}) for high thrust operations; second, cold fire mode (19 mN thrust and 47 s I_{sp}) for low thrust (attitude control) operations. Table 6.4 shows a summary of the surveyed resistojets and their performance parameters. None of these resistojets have a heritage of flying on a CubeSat mission, though one of them has been utilized on a bigger spacecraft (Coxhill and Gibbon, 2005; Hejmanowski *et al.*, 2015).

Table 6.4. Summary of Resistojet Propulsion Systems

Company/Institution with Location	Engine	Thrust (mN)	Power (W)	I_{sp} (s)	Propellant
Busek, Natick, MA, USA	AMR	10	15	150	R134a, R236fa
CU Aerospace, Champaign, IL, USA, and VACCO Industries Inc., Huntsville, AL, USA	CHIPS	30	30	82	R134a, R236fa
CU Aerospace, Champaign, IL, USA, and VACCO Industries Inc., Huntsville, AL, USA	PUC	5.4	15	65	SO_2
University of Southern California, Los Angeles, CA, USA	FMMR	0.129	–	79.2	Water
SSTL, Guildford, UK	LPR	18	30	48	Xe

6.6. RADIO-FREQUENCY ION THRUSTER (RIT)

6.6.1. Operating Principle

Radiofrequency ion thrusters belong to a subset of gridded ion thrusters that produce thrust by accelerating the ionized propellant (plasma) via an electrostatic grid. Electron bombardment and microwave thrusters are some additional gridded ion thrusters. In RITs, the stored propellant is let into the discharge chamber where it is ionized (and becomes plasma) using radio frequency (RF) power (from RF coils). The ionized propellant is later extracted (from the discharge chamber) and accelerated by a series of grids (ion optics) called screen and accelerator grids. The screen grid extracts propellant cations (for instance, Xe^+, Kr^+ ions) from the ionized plasma and directs them downstream to the accelerating grid. Bigger RF ion engines that are utilized on larger spacecraft also have a third grid called the decelerator grid. Nevertheless, it is typically not used in CubeSat propulsion systems. A neutralizer cathode, existing on the exterior of the thruster in all ion engines, delivers electrons to neutralize the ionized propellant that is released from the thruster.

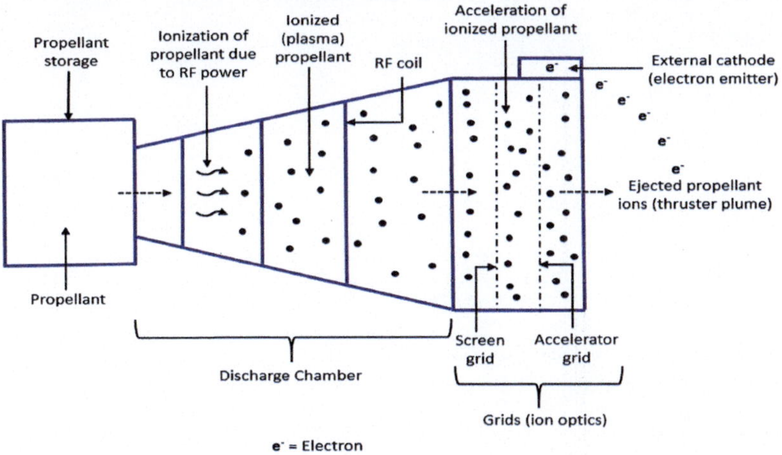

Figure 6.5. *Schematic of a radio frequency (RF) ion propulsion system.*

The specific impulse of a gridded thruster can be altered by varying the voltage that is provided to the accelerating grids (Goebel and Katz, 2008). Electron bombardment and microwave thrusters are additional types of gridded ion thrusters where the ionization happens because of electron bombardment with the neutral propellant and microwave power, respectively. The schematic of an RF Ion propulsion system is displayed in Figure 6.5 that includes the propellant storage, RF coil, discharge chamber, grids (screen and accelerator), and a neutralizing (external) cathode. Ion thrusters are characterized by high thruster efficiency (60% to >80%) causing high specific impulse (from 2000 s to over 10,000 s); though, they have been overwhelmed with problems that are caused by cathode wear and contamination over elongated usage. Several types and compositions of contamination are elucidated in the succeeding sub-section.

The ion exhaust velocity (shown in Eqn. (12)) and thrust (shown in Eqn. (13)) are both functions of the charge of propellant ion (q), the mass of propellant ion (m_{ion}), and ion accelerating voltage (V_i).

Ion engines utilize heavier elements (elements with higher atomic mass) as propellants since the thrust produced is proportionate to the mass of the ion (propellant).

Thrust, though, is also determined by the ion beam current (I_i). Specific impulse (shown in (14)) is a function of ion accelerating voltage and mass of ion (Carroll and Cardin, 2015). The performance factors of ion engines are clarified below with their mathematical equations (Parker, 2016):

$$v_{e_i} = \sqrt{\frac{2qV_i}{m_{ion}}} \tag{12}$$

$$\tau = \sqrt{\frac{2m_{ion}V_i}{q}} I_i \tag{13}$$

$$I_{sp} = 1.417 \times 10^3 \gamma_c \eta_m \sqrt{\frac{V_i}{m_{ion}}}, \tag{14}$$

Where; η_m is the thruster mass utilization efficiency; and γ_c is the total thrust correction factor.

6.6.2. Design Considerations and Technologies

Ion thruster operation might produce several interactions between the thruster and the spacecraft instruments, i.e., ionized propellant (plasma), contamination, and field interactions.

Plasma and contamination interactions consist of two types of efflux: propellant efflux comprises propellant ions, neutralizing electrons, non-ionized propellant, and a low-energy charge-exchange plasma; non-propellant efflux composed of material sputtered from the thruster components and the neutralizer because of ion bombardment. The field interactions are due to the RF field, electrostatic accelerators, and the interaction of the plasma plume with the ambient (space) environment. Contamination effects experienced can be diminished by the use of inert propellants like xenon and krypton, yet, it still leaves out the issues due to the plasma interactions (Ito *et al.*, 2007).

Hollow cathode tubes are frequently used in electron-bombardment ion engines to deliver electrons for the neutralization of the ion beam. The hollow cathode assembly comprises primarily of the cathode tube that has an insert (electron emitter), orifice plate (present downstream of thruster) that assists the flow of exhaust plume, heater that increases insert temperature, and a keeper electrode.

Key functions of a keeper electrode are to aid in turning on the cathode discharge, to preserve the cathode temperature and operation when discharge or beam current is disturbed provisionally, and to guard the cathode orifice plate and external heater from ion bombardment. It is of excessive significance to study and evaluate the wear of discharge cathodes as their failure is considered to be one of the foremost life-limiting mechanisms of ion thrusters.

The subsequent wear processes are usually seen in discharge cathode assembly: cathode orifice plate, failure of the heater, and keeper electrode (Yamamoto *et al.*, 2005).

A novel field of materials science called solid-state Ionics (SSIs) can propose solutions to overwhelmed problems with the creation and transfer of ions to their extraction sites deprived of problems of plasma discharge chamber and all of its crucial components.

SSIs deal with the theory, preparation, characterization, and application of solids that support ionic conduction. SSI conductors are presently the basic elements in the oxygen sensors used in automobile exhaust systems, solid oxide fuel cells, lithium-ion batteries, electrochromic windows, and in some superconductors. Table 6.5 shows the summary of the surveyed RF ion thrusters together with their performance factors (Zhurin *et al.,* 1999).

Not any of these ion thrusters have a space heritage, whereas one of them is scheduled to fly on the *Lunar IceCube* mission in 2018.

Table 6.5. Summary of RF Ion Propulsion Systems

Company/Institution with Location	Engine	I_{sp} (s)	Thrust (mN)	Power (W)	Propellant
Busek, Natick, MA, USA	BIT-3	2500	1.15	75	Iodine
Airbus, Lampoldshausen, Germany	RIT 10 EVO	>1900, >3000, >3200	5, 15, 25	145	Xe
Busek, Natick, MA, USA	BIT-1	2150–3200	0.1–0.18	28	Xe, Iodine
Airbus, Lampoldshausen, Germany	RIT-μX	300–3000	0.05–0.5	<50	Xe

6.7. HALL EFFECT PROPULSION/HALL THRUSTERS

6.7.1. Operating Principle

Hall Thrusters are electrostatic devices that produce thrust by initially ionizing and then accelerating the propellant in mutually perpendicular electric and magnetic fields. These thrusters' function on the principle of the recognized Hall Effect that addresses the following: when an electric current is provided to a conductive material (propellant) located in mutually perpendicular electric and magnetic fields, a potential difference is created which is perpendicular to the applied electric and magnetic fields. The schematic of a Hall thruster is displayed in Figure 6.6 that comprises discharge channel, external cathode, propellant storage, anodes, and the magnetic field generator. The supplied magnetic field is radial, while the accelerating electric field (acting from anode towards the cathode) is axial. Note that Hall thrusters, dissimilar to gridded ion thrusters, do not have the grid system (series of grids), as a substitute, the grids are interchanged with a strong magnetic field perpendicular to the flow of ions. This magnetic field lessens the mobility of electrons approaching from the external cathode, thus detaining their flow to the anode in the accelerating electric field. Hall thrusters have numerous beneficial features such as high specific impulse (higher than most systems exception engines), higher thrust density, and easiness in design (when associated with gridded ion engines owing to shortage of accelerator grids). Nonetheless, they also face certain challenges with wearing away magnetic circuitry because of discharge plasma and lower efficiency (6–30% at 0.1–0.2 kW and 50% at 1 kW) (Kieckhafer and King, 2007).

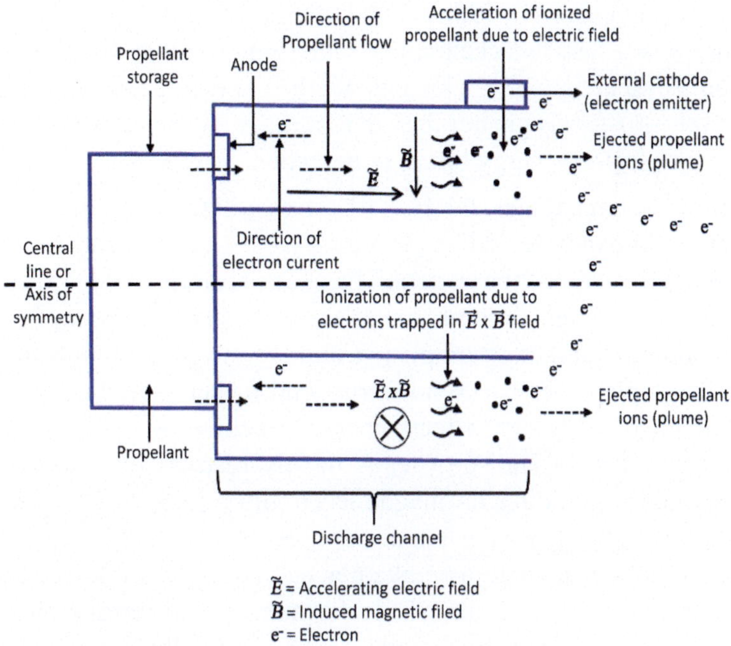

Figure 6.6. *Schematic of a Hall effect propulsion system.*

The performance factors for Hall thrusters such as ion exit velocity (shown in Eqn. (12), thrust (shown in Eqn. (13), and specific impulse (shown in Eqn. (14) are the same as the ones for RITs.

6.7.2. Design Considerations and Technologies

Remember that, in a distinctive Hall thruster, magnetic field (B˜) is applied across an accelerating electrical discharge (E˜) permitting to trap the electrons in the *Hall effect (*E˜ × B˜*) direction*. The anode is placed at the base of the discharge channel, and assists as the basis of the neutral propellant. An external cathode is located outside the discharge channel and delivers electrons that move towards the anode across the radial magnetic field. When the electrons pass in the magnetic field, they spiral around the thruster axis in the (E˜ × B˜) direction, and their contact with the incoming propellant results in the ionization of the propellant (Hillier *et al*., 2011).

Hall thrusters are generally categorized into two types: magnetic layer and anode layer thrusters. The magnetic layer thrusters have uninterrupted and prolonged acceleration zones for adequate ionization and stability. They also possess a ceramic wall, and their acceleration channel length is extended to the channel width. Alternatively, Hall thrusters with an anode layer comprise a narrow acceleration zone (length of the discharge channel is shorter compared to the channel width). The electron temperature of anode layer thrusters is greater than that of magnetic layer thrusters because of the lower electron energy losses (Cheng *et al*., 2008).

Comparable to ion engines, the Hall thrusters utilize heavy elements as propellants, for example, iodine (I), xenon (Xe), krypton (Kr), bismuth (Bi), and argon (Ar). Of these, xenon has been preferred due to its lower ionization energy, higher atomic mass, and easy storage. Though, it is affluent to purchase and to accomplish ground tests with xenon. Numerous inexpensive replacements to xenon exist, but further experiments have to be conducted to verify their effectiveness (Mikellides *et al.,* 2014).

The lifespan of a Hall thruster is mostly restricted by the erosion of the components shielding its magnetic circuitry from discharged plasma (ionized propellant). Once the magnetic poles are uncovered over time, more degradation or overheating may take place, affecting the nominal magnetic field and in that way the thruster's performance. To determine the lifetime of a Hall thruster, apart from the traditional long-duration qualification tests, shorter duration experiments are also executed so that the erosion behavior can be investigated facilitating the extrapolation of the thruster lifetime (Biagioni *et al.,* 2003).

As compared to gridded ion thrusters where ion beam can be appropriately controlled, it is further problematic to control the same in Hall thrusters causing the wall erosion. This problem has troubled Hall thrusters for various decades since they were presented for larger spacecraft. Wall erosion is principally initiated when the ions are driven in the direction of the wall material as a consequence of the prominent parallel component of the electric field and the high electron temperature. In recent times, a novel technique called magnetic shielding (MS) was recommended that could effectively eradicate wall erosion in Hall thrusters. It is to be noted that, the magnetic and electric fields that are thought to be mutually perpendicular are not so when under electron pressure (Reed *et al.,* 2006). When the walls are magnetically protected, the electric field component parallel to the wall is approximately removed, resultant in the reduction of ion bombardment on the walls. For example, when a magnetically shielded Boron Nitrate wall was utilized at an ion threshold energy of 25 V, the computed wall erosion rate was found to be almost 600 times lower at the inner wall (wall closer to central line) than when unshielded. The outer wall instead was found to experience zero erosion (Dannenmayer and Mazouffre, 2011).

REFERENCES

1. Ahmed, Z., Gimelshein, S. F., & Ketsdever, A. D., (2006). Numerical analysis of free-molecule microresistojet performance. *J. Propuls. Power, 22*, 749–756.
2. Anis, A., (2012). Cold gas propulsion system-an ideal choice for remote sensing small satellites. In: *Remote Sensing-Advanced Techniques and Platforms* (Vol. 1, pp. 1–33). InTech: Rijeka, Croatia.
3. Barnhart, D. J., Vladimirova, T., & Sweeting, M. N., (2007). Very-small-satellite design for distributed space missions. *Spacecr. Rocket, 44*, 1294–1306.
4. Bauer, H., (1963). Stability boundaries of liquid-propelled space vehicles with sloshing. *AIAA J., 1*, 1583–1589.
5. Biagioni, L., Saverdi, M., & Andrenucci, M., (2003). Scaling and performance prediction of hall effect thrusters. In: *Proceedings of the 39th AIAA/ASME/SAE/ASEE Joint Propulsion Conference and Exhibit* (pp. 1–33) Huntsville, AL, USA.
6. Bonin, G., Roth, N., Armitage, S., Newman, J., Risi, B., & Zee, R. E., (2015). CanX-4 and CanX-5 precision formation flight: Mission accomplished! In: *Proceedings of the 29th Annual AIAA/USA Conference on Small Satellites* (pp. 1–30). Logan, UT, USA.
7. Bowen, J., Villa, M., & Williams, A., (2015). CubeSat based rendezvous, proximity operations, and docking in the CPOD mission. In: *Proceedings of the 29th Annual AIAA/USA Conference on Small Satellites* (Vol. 29, p. 9). Logan, UT, USA.
8. Bradford, A., Davies, P., Liddle, D., Paffett, J., Sweeting, M., Tondryk, W., Gatti, G., & Garutti, A., (2006). The GIOVE-A small navigation mission. In: *Proceedings of the Small Satellite Conference* (Vol. 2006, pp. 1–39). Logan, UT, USA.
9. Carpenter, C. B., Schmuland, D., Overly, J., & Masse, R., (2013). CubeSat modular propulsion systems product line development status and mission applications. In: *Proceedings of the 49th AIAA/ASME/SAE/ASEE Joint Propulsion Conference* (Vol. 49, pp. 1–32). San Jose, CA, USA.
10. Carroll, D. L. J. M., & Cardin, R. L. B., (2015). Propulsion unit for CubeSats (PUC). In: *Proceedings of the 62nd JANNAF Propulsion Meeting (7th Spacecraft Propulsion)* (pp. 10–45). Nashville, TN, USA.

11. Cheng, S. Y., & Martinez-Sanchez, M., (2008). Hybrid particle-in-cell erosion modeling of two hall thrusters. *J. Propuls. Power, 24*, 987–998.
12. Chianese, S. G., & Micci, M. M., (2006). Microwave electrothermal thruster chamber temperature measurements and performance calculations. *J. Propuls. Power, 22*, 31.
13. Coxhill, I., & Gibbon, D. A., (2005). Xenon resistojet propulsion system for microsatellites. In: *Proceedings of the 41st AIAA/ASME/SAE/ASEE Joint Propulsion Conference and Exhibit* (pp. 1–28). Tucson, AZ, USA.
14. Dannenmayer, K., & Mazouffre, S., (2011). Elementary scaling relations for hall effect thrusters. *J. Propuls. Power, 27*, 236.
15. David, A. O., & Knoll, A. K., (2017). Experimental demonstration of an aluminum-fueled propulsion system for CubeSat applications. *J. Propuls. Power, 33*, 1320–1324.
16. Davies, P., Whittaker, P., Bird, R., Gomes, L., Stern, B., Sweeting, M., Cohen, M., & Hall, D., (2012). NovaSAR-bringing radar capability to the disaster monitoring constellation. In: *Proceedings of the Small Satellite Conference* (pp. 1–37) Logan, UT, USA.
17. Friedhoff, P., Hawkins, A., Carrico, J., Dyer, J., & Anflo, K., (2017). On-orbit operation and performance of ammonium dinitramide (ADN) based high performance green propulsion (HPGP) systems. In: *Proceedings of the 53rd AIAA/SAE/ASEE Joint Propulsion Conference* (p. 4673). Atlanta, GA, USA.
18. Frisbee, R. H., (2003). Advanced space propulsion for the 21st century. *J. Propuls. Power, 19*, 1129–1154.
19. Gibbon, D. A., (2010). Review of the use of butane as a low-cost propellant. In: *Proceedings of the 2010 Space Propulsion* (Vol. 2010, p. 11). San Sebastian, Spain.
20. Goebel, D. M., & Katz, I., (2008). *Fundamentals of Electric Propulsion: Ion and Hall Thrusters* (Vol. 1, pp. 6–7, 22, 26, 148, 189, 192, 243–244, 325, 327). Jet Propulsion Laboratory (JPL): Pasadena, CA, USA.
21. Hejmanowski, N. J. C. A., & Woodruff, R. B., (2015). CubeSat high impulse propulsion system (CHIPS). In: *Proceedings of the 62nd JANNAF Propulsion Meeting (7th Spacecraft Propulsion)* (pp. 1–34). Nashville, TN, USA.
22. Hevner, R., Holemans, W., Puig-Suari, J., & Twiggs, R., (2011). An advanced standard for CubeSats. In: *Proceedings of the 25th Annual*

AIAA/USU Conference on Small Satellites (Vol. 25, pp. 1–40). Logan, UT, USA.

23. Hillier, A., Branam, R., Huffman, R., Szabo, J., & Paintal, S., (2011). High thrust density propellants in hall thrusters. In: *Proceedings of the 49th AIAA Aerospace Sciences Meeting Including the New Horizons Forum and Aerospace Exposition* (p. 524). Orlando, FL, USA.
24. Ito, T., Gascon, N., Crawford, W. S., & Cappelli, M. A., (2007). Experimental characterization of a micro-hall thruster. *J. Propuls. Power, 23*, 1068–1074.
25. James, K., Moser, T., Conley, A., Slostad, J., & Hoyt, R., (2015). Performance characterization of the HYDROS water electrolysis thruster. In: *Proceedings of the 29th Annual AIAA/USU Conference on Small Satellites* (Vol. 29, pp. 1–20). Logan, UT, USA.
26. Ketsdever, A. D., Wadsworth, D. C., & Muntz, E., (2001). Gas-surface interaction model influence on predicted performance of microelectromechanical system resistojet. *J. Thermo Phys. Heat Transf., 15*, 302–307.
27. Kieckhafer, A., & King, L. B., (2007). Energetics of propellant options for high-power hall thrusters. *J. Propuls. Power, 23*, 21–26.
28. Klesh, A., Baker, J., Castillo-Rogez, J., Halatek, L., Murphy, N., Raymond, C., Sherwood, B., et al., (2013). Inspire: Interplanetary nano spacecraft pathfinder in relevant environment. In: *Proceedings of the AIAA SPACE 2013 Conference and Exposition* (Vol. 2013, pp. 1–39). San Diego, CA, USA.
29. Kolmas, J., Banazadeh, P., Koenig, A. W., Macintosh, B., & D'Amico, S., (2016). System design of a miniaturized distributed occulter/telescope for direct imaging of star vicinity. In: *Proceedings of the 2016 IEEE Aerospace, Big Sky* (Vol. 2016, pp. 1–11). MT, USA.
30. Kolosa, D., Spangelo, S., Lemmer, K., & Hudson, J., (2014). Mission analysis for a micro rf ion thruster for CubeSat orbital maneuvers. In: *Proceedings of the Joint Propulsion Conference* (pp. 1–13). Cleveland, OH, USA.
31. Lee, R. H., Bauer, A., Killingsworth, M. D., Lilly, T., Duncan, J., & Ketsdever, A., (2008). Free-molecule-microresistojet performance using water propellant for nanosatellite applications. *J. Spacecr. Rocket, 45*, 264–269.

32. Legge, R. S., Clements, E. B., & Shabshelowitz, A., (2017). Enabling microsatellite maneuverability: A survey of microsatellite propulsion technologies. In: *Proceedings of the 2017 IEEE MTT-S International Microwave Symposium (IMS)* (pp. 229–232). Honolulu, HI, USA.
33. Lemmer, K., (2017). Propulsion for CubeSats. *Acta Astronaut., 134*, 231–243.
34. Ley, W., Wittmann, K., & Hallmann, W., (2009). *Handbook of Space Technology* (Vol. 22, p. 154). John Wiley & Sons: Hoboken, NJ, USA.
35. Manzoni, G., & Brama, Y. L., (2015). CubeSat micro propulsion characterization in low earth orbit. In: *Proceedings of the 29th Annual AIAA/USA Conference on Small Satellites* (Vol. 29, pp. 1–35). Logan, UT, USA.
36. Martinez-Sanchez, M., & Pollard, J. E., (1998). Spacecraft electric propulsion: An overview. *J. Propuls. Power, 14*, 688–699.
37. Matticari, G., Noci, G., Siciliano, P., Colangelo, G., & Schmidt, R., (2006). Cold gas micro propulsion prototype for very fine spacecraft attitude/position control. In: *Proceedings of the AIAA/ASME/SAE/ASEE 42nd Joint Propulsion Conference* (pp. 5378–5390). Sacramento, CA, USA.
38. Mikellides, I. G., Katz, I., Hofer, R. R., & Goebel, D. M., (2014). Magnetic shielding of a laboratory hall thruster. I. Theory and validation. *J. Appl. Phys., 115*, 043303.
39. Mission Design Division, (2015). *Small Spacecraft Technology State of the Art* (pp. 40–60) NASA: Moffett Field, CA, USA, Number: NASA/TP-2015-216648/REV1.
40. Mueller, J., Ziemer, J., Hofer, R., Wirz, R., & O'Donnell, T., (2008). A survey of micro-thrust propulsion options for micro spacecraft and formation flying missions. In: *Proceedings of the 5th Annual CubeSat Developers Workshop* (pp. 1–27). San Luis Obispo, CA, USA.
41. Nervold, A. K., Berk, J., Straub, J., & Whalen, D., (2016). A pathway to small satellite market growth. *Adv. Aerosp. Sci. Technol., 1*, 14.
42. Nicholas, A., Finne, T., Galysh, I., Mai, A., Yen, J., Sawka, W., Ransdell, J., & Williams, S., (2013). Spin sat mission overview. In: *Proceedings of the 27th Annual AIAA/USA Conference on Small Satellites* (pp. 1–38). Logan, UT, USA.
43. Parker, K. I., (2016). State-of-the-art for small satellite propulsion systems. In: *Proceedings of the 2016 Biennial Aerospace Systems*

Conference of the National Society of Black Engineers (NSBE) (pp. 5–37). Arlington, VA, USA.

44. Persson, S., Jacobsson, B., & Gill, E., (2005). PRISMA-demonstration mission for advanced rendezvous and formation flying technologies and sensors. In: *Proceedings of the 56th International Astronautical Congress* (Vol. 56, pp. 1–30). Fukuoka, Japan.
45. Reed, G. D., Hargus, W. A., & VanGilder, D. B., (2006). *Comparison of Numerical and Experimental Near-Field Plasma Properties of the BHT-200-X3 Hall Thruster (Preprint)* (pp. 1–33). Technical Report; Air Force Research Lab, Edwards Air Force Base: Kern, CA, USA.
46. Reid, M. R., Scharfe, D. B., Saleem, F. A., & Webb, R. N., (2013). Preheating cold gas thruster flow through a thermal energy storage conversion system. *J. Propuls. Power, 29*, 1488–1492.
47. Robin, M., Brogan, T., & Cardiff, E., (2008). An ammonia microresistojet (MRJ) for microsatellites. In: *Proceedings of the 44th AIAA/ASME/SAE/ASEE Joint Propulsion Conference and Exhibit* (pp. 1–52). Hartford, CT, USA.
48. Roscoe, C. W., Westphal, J. J., Griesbach, J. D., & Schaub, H., (2015). Formation establishment and reconfiguration using differential elements in J2-perturbed orbits. *J. Guid. Control Dyn., 38*, 1725–1740.
49. Rossi, C., Larangot, B., Pham, P. Q., Briand, D., De Rooij, N. F., Puig-Vidal, M., & Samitier, J., (2006). Solid propellant micro thrusters on silicon: Design, modeling, fabrication, and testing. *J. Microelectron. Syst., 15*, 1805–1815.
50. Sackheim, R. L., & Masse, R. K., (2014). Green propulsion advancement: Challenging the maturity of monopropellant hydrazine. *J. Propuls. Power, 30*, 265–276.
51. Sathiyanathan, K., Lee, R., Chesser, H., Dubois, C., Stowe, R., Farinaccio, R., & Ringuette, S., (2011). Solid propellant microthruster design for nanosatellite applications. *J. Propuls. Power, 27*, 1288–1294.
52. Sawka, W. N., & McPherson, M., (2013). Electrical solid propellants: A safe, micro to macro propulsion technology. In: *Proceedings of the 49th AIAA/ASME/SAE/ASEE Joint Propulsion Conference* (pp. 11–42). San Jose, CA, USA.
53. Schmuland, D. T., Carpenter, C., & Masse, R. K., (2012). Mission applications of the MRS-142 CubeSat high-impulse adaptable

monopropellant propulsion system (CHAMPS). In: *Proceedings of the 48th AIAA/ASME/SAE/ASEE Joint Propulsion Conference and Exhibit* (Vol. 48, pp. 1–34). Atlanta, GA, USA.

54. Schmuland, D., Masse, R., & Sota, C., (2011). Hydrazine propulsion module for CubeSats. In: *Proceedings of the 25th Annual AIAA/USU Conference on Small Satellites* (Vol. 25, pp. 1–29). Logan, UT, USA.
55. Schoolcraft, J., Klesh, A., & Werne, T., (2017). MarCO: Interplanetary mission development on a CubeSat scale. In *Space Operations: Contributions from the Global Community* (pp. 221–231). Springer: Daejeon, Korea.
56. Skuhersky, M., Balestrero, M. L., Wilde, M., & Brett, C., (2017). CERBERUS: Prototype for an agile inspection and servicing satellite using thrust-vectoring cold-gas propulsion. In: *Proceedings of the AIAA SPACE and Astronautics Forum and Exposition* (p. 5116). Orlando, FL, USA.
57. Slough, J., Andreason, S., Ziemba, T., & Ewing, J., (2005). Micro-discharge micro-thruster. In: *Proceedings of the 41st AIAA/ASME/SAE/ASEE Joint Propulsion Conference and Exhibit* (Vol. 4047, pp. 1–44). Tucson, AZ, USA.
58. Spores, R. A., (2015). GPIM AF-M315E propulsion system. In: *Proceedings of the 51st AIAA/SAE/ASEE Joint Propulsion Conference Number AIAA* (Vol. 51, p. 3753). Orlando, FL, USA.
59. Spores, R. A., Masse, R., Kimbrel, S., & McLean, C., (2013). GPIM AF-M315E propulsion system. In: *Proceedings of the 49th AIAA Joint Propulsion Conference* (Vol. 49, pp. 1–30). San Jose, CA, USA.
60. Stratton, J., (2004). The use of the AEROJET MR-103H thruster on the new horizons mission to Pluto. In: *Proceedings of the 55th International Astronautical Congress 2004* (Vol. 30, p. 30) Vancouver, BC, Canada.
61. Sutton, G. P., & Biblarz, O., (2001). *Rocket Propulsion Elements* (Vol. 7, pp. 31–32, 34, 52, 133, 253, 260–261, 315, 417, 428, 608). John Wiley & Sons, Inc.: Hoboken, NJ, USA.
62. Thrasher, J., Williams, S., Takahashi, P., & Sousa, J., (2016). Pulsed plasma thruster development using a novel HAN-based green electric monopropellant. In: *Proceedings of the 52nd AIAA/SAE/ASEE Joint Propulsion Conference* (p. 4846). Salt Lake City, UT, USA.
63. Toorian, A., Blundell, E., Suari, J. P., & Twiggs, R., (2005). CubeSats as responsive satellites. *Aerosp. Eng., 805*, 756–6479.

64. Tsay, M., Feng, C., & Zwahlen, J., (2017). System-level demonstration of Busek's 1U CubeSat green propulsion module "AMAC." In: *Proceedings of the 53rd AIAA/SAE/ASEE Joint Propulsion Conference* (p. 4946). Atlanta, GA, USA.
65. Tsay, M., Zwahlen, J., Lafko, D., Feng, C., & Robin, M., (2016). Complete EM system development for Busek's 1U CubeSat green propulsion module. In: *Proceedings of the 52nd AIAA/SAE/ASEE Joint Propulsion Conference* (Vol. 52, pp. 1–32). Salt Lake City, UT, USA.
66. Underwood, C. I., Richardson, G., & Savignol, J., (2003). In-orbit results from the SNAP-1 nanosatellite and its future potential. *Philos. Trans. R. Soc. Lond. A Math. Phys. Eng. Sci., 361*, 199–203.
67. Whitmore, S. A., (2017). Three-dimensional printing of "green" fuels for low-cost small spacecraft propulsion systems. *J. Spacecr. Rocket, 6*, 1–14.
68. Woellert, K., Ehrenfreund, P., Ricco, A. J., & Hertzfeld, H., (2011). CubeSats: Cost-effective science and technology platforms for emerging and developing nations. *Adv. Space Res., 47*, 663–684.
69. Wu, S., Mu, Z., Chen, W., Rodrigues, P., Mendes, R., & Alminde, L., (2014). TW-1: A CubeSat constellation for space networking experiments. In: *Proceedings of the 6th European CubeSat Symposium* (Vol. 2014, pp. 1–25). Estavayer-le-Lac, Switzerland.
70. Yamamoto, N., Komurasaki, K., & Arakawa, Y., (2005). Discharge current oscillation in hall thrusters. *J. Propuls. Power, 21*, 870–876.
71. Zhang, K., Chou, S., & Ang, S. S., (2004). Development of a solid propellant micro thruster with chamber and nozzle etched on a wafer surface. *J. Micromech. Microeng., 14*, 785.
72. Zhurin, V., Kaufman, H., & Robinson, R., (1999). Physics of closed drift thrusters. *Plasma Sources Sci. Technol., 8*, R1.
73. Zondervan, K., Fuller, J., & Rowen, E., (2014). CubeSat solid rocket motor propulsion systems providing delta-vs greater than 500 m/s. In: *Proceedings of the 28th Annual AIAA/USA Conference on Small Satellites* (pp. 1–30). Logan, UT, USA.

Chapter

7

SMALL SATELLITES MISSIONS

CONTENT

7.1. Introduction..156
7.2. Classification 160
7.3. Ten-Koh Project...172
References 179

7.1. INTRODUCTION

A miniaturized satellite, small satellite, or simply small-sat is the satellite having low size and mass, generally less than 500 kg. Whereas all such kinds of satellites can be termed as small, diverse classifications are utilized to classify them centered on mass. The satellites can be made small to decrease the large economic expense of launch vehicles and expenses linked with construction. Small satellites, particularly in large numbers, might be more beneficial than fewer, larger satellites for some objectives-for instance, radio relay and collection of scientific data. Technical encounters in the production of small satellites might include the shortage of adequate power storage or room for the propulsion system (Sihver *et al.*, 2016).

One method of categorizing satellites is on the basis of the satellite's in-orbit mass. Centered on this principle, satellites are usually classified as femto, pico nano, micro, mini, medium, or large (Dandumont et al., 2020).

Femto, pico, nano, micro, and minisatellites are mutually categorized as *miniature or small satellites*. Table 7.1 displays the cataloging of satellites centered on wet mass, i.e., the satellite's mass including fuel. Today the commercial space sector is acknowledged by the geostationary communications satellite systems. One of the main driving forces accountable for the applicability of small satellites with the passage of years normally has been the necessity to empower missions that the larger satellites could not have accomplished. These comprised constellations for lower data rate communications, university-linked research, and in-orbit examination of the larger satellites. Other assisting aspects have been the necessity for cheaper and smaller launch vehicles as contrary to the larger rockets proficient in generating greater thrust and therefore greater financial price for heavier and larger satellites (Woerd and Wernand, 2015). Also, the smaller satellites can usually be launched in many numbers and as the piggybacks utilizing excess capacity on the larger launch vehicles. These satellites have a lower manufacturing cost, simplicity of mass manufacturing, and faster building times, making these satellites the perfect test bed for novel technologies. The small satellites are not the exclusive perquisite of defense departments and the other main R&D organizations. These satellites are also a big attraction for universities and the commercial industry. According to an approximation, near to 1000 satellites have been placed into orbit between 2000 and 2020 in the group of small satellites, comprising femto, pico, nano, micro, and mini-satellites (Hussmann *et al.*, 2006; Paziewski *et al.*, 2018).

Table 7.1. Grouping of Satellites Centered on Wet Mass

Satellite Class	Wet Mass (kg)
Femto	<0.1
Pico	0.1–1
Nano	1–10
Micro	10–100
Mini	100–500
Medium	500–1000
Large	>1000

7.1.1. The Rationale for Miniaturized/Small Satellites

One rationale for reducing the size and mass of satellites is to decrease the price; heavier satellites need larger rockets having greater thrust that brings with it greater price to finance. On the contrary, lighter, and smaller satellites need cheaper and smaller launch vehicles and at times can be launched in greater numbers. They can be launched 'piggyback,' utilizing excess capacity on the larger launch vehicles. Smaller satellites permit inexpensive designs and the simplicity of mass production (Suparta, 2014).

Another main reason for evolving smaller satellites is the chance to empower missions that the larger satellite could not accomplish, like (Chen et al., 2012):

- Constellations for the low data rate communications;
- In-orbit examination of larger satellites;
- Using formations to collect data from several points;
- Testing or qualifying novel hardware before utilizing it on the more expensive spacecraft;
- University-associated research.

Miniaturized satellites and their uses have opened the room for several countries and their non-governmental and governmental organizations, comprising universities, research, and education institutes, and the private industry, with inadequate funds for the activities related to space to join in investigation and peaceful uses of the outer space and in order to become developers of the space technology. Satellites might be classified into diverse categories centered on their mass (for instance, mini satellites less than 100 kilograms, nanosatellites less than 10 kilograms, picosatellites less than 1 kilogram, Femto satellites less than 0.1 kilograms). However, to date, there is no consent or universally acknowledged standard on the definition of the small satellite. A small satellite is not essentially small physically as it might possess deployable structures, it is not essentially low-weight, and neither does the satellite need to be less intricate or less capable as compared to the satellite that is not well-thought-out to be small. Typical features of the small satellite missions comprise (Cucinotta et al., 2020):

- Practically short development times;
- Comparatively small development teams;
- Modest development and the testing infrastructure necessities;
- Reasonable development and operation expenses for the developers, "faster, smaller, and cheaper."

Some other features observed in the small satellite missions include (Miyake et al., 2017):

- They comprise actors' novel to space activities primarily non-governmental actors such as private companies and academic institutions.
- For several reasons, very often because of inexperience or unacquaintedness with the international and national regulatory framework, they are not always conducted in full compliance with international regulations, obligations, and appropriate voluntary guidelines.
- They have raised worries about deteriorating the situation of space debris.

For the liftoff and function of satellites, several requirements under international law prevail. These comprise: (i) notification and record of the RF frequencies utilized by the satellite at ITU (International Telecommunications Union); (ii) consideration of measures for the space debris mitigation in the design and function of the satellite; (iii) registration of the satellite with Secretary-General of UN (United Nations). Currently, a regulatory or legal definition of the small satellite does not exist. Under the UN treaties, resolutions, and principles linking to the international law regarding space, the term space object refers to launch vehicles, satellites, and their parts.

Space radiation disturbs satellites by presenting anomalies like SEE (single event effects), component degradation because of ionizing radiation dose, and internal and surface charging. Comprehending the environment of radiation is, thus, significant to develop satellites that can endure the feasible anomalies. The sources of space radiation comprise GCR (galactic cosmic rays), SEP (solar energetic particle) events, particles with high energy trapped in the magnetic field of Earth, and the constant radiation background. Additionally, the LEO (low-Earth orbit) region where most of the satellites exist is subject to unidentified mechanisms that give it random energy inconsistency in the particle spectra. The variability is mainly poorly comprehended for electrons, which normally can appear with higher energies than anticipated (Kirby *et al.,* 2012).

The concerns of the existence of electrons with high-energy, ions, and protons for spacecraft designers differ depending on every mission design, class, and epoch. The design of spacecraft must consider the effects of ionizing radiation, along with discharging and charging effects on the surfaces of the satellite. For manned missions, the duration of the mission and the linked life support systems must be tuned to consider this unpredictability of the particle populations (Stratton *et al*., 2013).

In current years, diverse missions have been put into orbit to discover the near-Earth region to give direct measurements of the charged particles, magnetosphere, and plasma. Missions like RBSP (Van Allen Probes), MMS, THEMIS from NASA, JAXA's ERG satellite, Swarm, and Proba-2 from the European Space Agency (ESA) have been put into particular orbits to explore space radiation around the planet Earth (Sharma and Curtis, 2005). Moreover, to the missions mentioned having launch masses of nearly hundreds of kg, devoted small satellites provide a significant counterpart to the measurements since they empower a wider, more broad view of the environment of space. This category of satellites leverages COTS (commercial off-the-shelf) components to bring time and cost savings at the price of an augmented risk of failure. The miniaturized-satellite Ten-Koh has been designed to demonstrate the possibility of providing measurements of space environment with such low-price platforms, along with giving readily usable data (Angelopoulos et al., 2013; Nakamura *et al.,* 2018).

7.1.2. History

The microsatellite and nanosatellite sections of the launch industry associated with the satellite have been growing quickly in recent years. The activity of development in the 1 to 50 kg range has been considerably surpassing that in the 50 to 100 kg range (Konecny, 2004).In the 1 to 50 kilograms range alone, less than 15 satellites were put into orbit annually in

2000–2005, 34 in the year 2006, then less than 30 liftoffs annually during 2007–2011. This escalated to 34 launches in 2012 and then 92 launches in 2013. European analyst Euro-consult projects around 500 small-sats being put into orbit in 2015 to 2019 with the market value projected at US$ 7.4 billion. By mid-2015, several more launch choices had become accessible for small-sats and rides as the secondary payloads had become greater in number and simple to schedule on very short notice.

7.2. CLASSIFICATION

Small satellites can normally be categorized into general small satellites, Femto-satellites, pico-satellites, nano-satellites, and micro-satellites (Buchen and DePasquale, 2014).

7.2.1. Small Satellites

The term small satellite, or at times minisatellite, generally refers to the artificial satellite having the wet mass (with fuel) between 100 and 500 kg, but in other practice, any satellite less than 500 kg.

Small satellite examples comprise Demeter, Parasol, Essaim, Picard, TARANIS, MICROSCOPE, ELISA, SMART-1, SSOT, Spirale-A & -B, and Star-link satellites (Messier, 2015).

Even though small-sats have conventionally been put into orbit as the secondary payloads on quite larger launch vehicles, various companies presently have developed or are developing launch vehicles explicitly targeted at the market of small-sat. Particularly, the paradigm of the secondary payload does not provide the specificity needed for various small satellites that possess exclusive launch-timing and orbital requirements.

Companies providing SmallSat launch vehicles comprise (Foust, 2015; Wisnarama, 2014):

- Rocket Lab's Electron (225 kilograms).

Companies planning to have SmallSat launch vehicles comprise:

- Astra's Rocket 3.0 (100 kilograms);
- Virgin Orbit's LauncherOne (500 kilograms).

7.2.2. Medium Satellites

The wet mass of these satellites varies in the range of 500 to 1000 kilograms. Medium satellites, even though simpler and smaller as compared to large satellites, utilize similar technologies as those utilized in large satellites. Numerous satellites developed for weather forecasting and remote sensing applications lie in the group of medium satellites (Izquierdo and Tristancho, 2011; Kil and Paxton, 2017).

7.2.3. Mini Satellites

The wet mass of these satellites lies in the range of 100 to 500 kg. Various satellites meant for military intelligence and surveillance, scientific studies, and satellites developed for earth observation and weather forecasting applications are all mini-satellites. Some instances of mini-satellites comprise SARAL, Jason-1, and Jason-2 satellites for the applications of remote sensing, the SMART-1 satellite having the aim of scientific studies, Electronic Intelligence by Satellite (ELISA-1-ELISA-4) and Systeme Pr` eparatoire Infra Rouge pour l'Alerte (SPIRALE) for military intelligence and the series of′ METEOSAT for worldwide weather forecasting. Jason-1 and Jason-2 satellites, correspondingly placed into orbit in 2001 and 2008, are the mutual project between CNES (France) and NASA (USA) and are meant to monitor worldwide climate forecasts and global ocean circulation, and to measure the ocean surface topography (Bouwmeester and Guo, 2010; Laštovička-Medin, 2016). The satellite with CityArgos and Altika SARAL is the altimetry technology mission by CNES and ISRO (India) put into orbit in 2013 from the Indian PSLV-C20. This mission is corresponding to the Jason-2 satellite. ELISA-1-ELISA-4 are the French military satellites. The program of ELISA is the demonstration system that covers the way for the intentional radar monitoring system. SPIRALE was put into orbit in 2009 and is meant to spot the ballistic missile flights in a boost phase by utilizing IR satellite imagery.

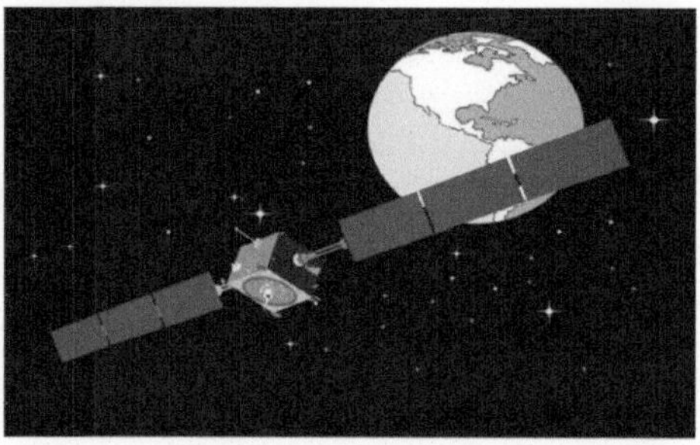

Figure 7.1. *SMART-1.*

Source: https://en.wikipedia.org/wiki/SMART-1.(Fig is not necessary)

SMART-1 satellite from ESA (Figure 7.1) was put into orbit in 2003 and is meant to orbit around the moon. SMART-1 had instruments for the lunar imaging, spotting of the chemical elements on the surface of the moon, spotting of the mineral spectra of pyroxene and olivine. The satellite was purposely made to bang into the surface of the moon in September 2006 (Tristancho and Gutierrez-Cabello, 2011; Jones, 2014).

7.2.4. Microsatellites

The term "microsat" or "microsatellite" generally refers to the artificial satellite having a wet mass between 10 and 100 kg. However, this is not an official tradition, and at times those terms can also refer to the satellites smaller than that, or larger than that. Sometimes, proposed designs or designs from some of the satellites of these kinds have microsatellites functioning together or in the formation. The general term "small-sat" or "small satellite" is also at times utilized, as is the "satlet" (Konecny, 2004; Buchen and DePasquale, 2014).

These satellites fall into the wet mass category of 10 to 100 kg. Several famous early satellites put into orbit during the 1950s–1970s belong to the category of microsatellite mainly due to the narrow launcher capacity obtainable during this period. Some of the common instances are Sputnik (1957); Telstar-1 (1962); Vanguard-1 (1958); Syncom-1 (1963); Apollo-P, and Early Bird (1965); and F1 (1971). With progress in technology empowering advanced payloads to be made into smaller volumes, there normally has been a new curiosity in a scientific commercial sector for microsatellites. The inclination started with the liftoff of UoSAT-1 in the year 1981, the 1st satellite to carry with it the microprocessor (Bonnici et al., 2019; Antonello et al., 2020). Current microsatellites normally carry with them an onboard computer (OBC) empowering these satellites to carry out the in-orbit programmable functions outside the range of a ground station. Some instances of the microsatellites put into orbit after the UoSAT-1 during the 1980s comprise the series of Cosmos, which makes part of the military-strategic communications constellations, the series of Iskra of the amateur RRS (radio relay satellites), the Japanese RRS Fuji-1 and the Rohini-3, which is meant for the remote sensing experiments. This inclination has continued over the 1990s–2000s. Some famous examples comprise Astrid-1 (1995) and Astrid-2 (1998), meant for the scientific studies, the FalconSat-1 (2000) for the technology demonstration in order to carry out an examination of an ion current collection in the wake of plasma, UNISAT-2 (2002) for the scientific research, atmospheric neutral density experiment

(ANDE, 2009) to measure composition and density of low earth orbit (LEO) atmosphere whereas being tracked from the ground, HAMSAT (2005) for the amateur radio communications, WNISAT (2012) for the applications of remote sensing applications and Near-Earth Object Surveillance Satellite (NEOSSat, 2013) (Figure 7.2) for noticing and tracking the asteroids that might create a threat to planet Earth and also for tracking orbital debris and satellites (Richie et al., 2007; Somov et al., 2014).

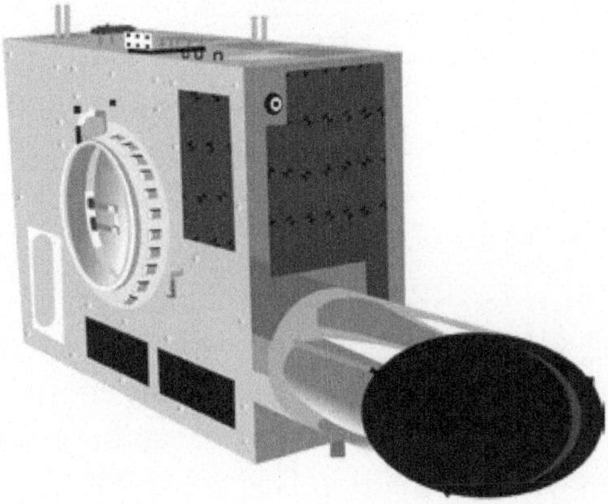

Figure 7.2. *NEOSSat.*

Source: https://directory.eoportal.org/web/eoportal/satellite-missions/n/neos-sat.

Examples: Astrid-1 and -2, along with the set of satellites presently announced for the *Launcher One* (below).

In the year 2018, the 2 Mars Cube One micro-sats—each having a mass of just 13.5 kg—became the 1st CubeSats to leave the orbit of Earth for utilization in interplanetary space. These satellites flew on the way to Mars along with the fruitful Mars *InSight* lander mission. The 2 microsats achieved the flyby of Mars in 2018 and continued communicating with the ground stations on the planet Earth through December. The 2 microsats went silent by January 2019 (Spector, 2020).

Several military-contractor and commercial companies are presently evolving microsatellite launch vehicles in order to execute the progressively targeted launch necessities of the microsatellites. Whereas microsatellites have been taken to space for several years as the secondary payloads onboard larger launchers, the paradigm of the secondary payload does not provide the specificity needed for numerous increasingly advanced small satellites that possess exclusive launch-timing and orbital requirements (Tristancho Martínez, 2010).

In the year 2012, Virgin Galactic publicized LauncherOne, the orbital launch vehicle developed to launch SmallSat as primary payloads of almost 100 kg into LEO. Various commercial customers contracted for launches, comprising GeoOptics, Spaceflight Industries, Planetary Resources, and Skybox Imaging. Both Sierra Nevada Space Systems and Surrey Satellite Technology are designing satellite buses enhanced to the design of launch vehicle LauncherOne (Werner, 2013). Virgin Galactic has usually been working since late 2008 on the concept of LauncherOne, and till 2015, is making it the larger part of the core business plan of Virgin as the program of Virgin human spaceflight has experienced various delays and the deadly accident in 2014 (Walker *et al.*, 2020).

In the year 2012, DARPA publicized that the program of Airborne Launch Assist Space Access would give them the microsat rocket booster for the SeeMe program that planned to release the constellation of 24 micro-satellites each having the one-meter imaging resolution (Graydon and Parks, 2020). The program was called off in December 2015.

The Boeing Small Launch Vehicle is the air-launched 3-stage-to-orbit concept launch vehicle meant to liftoff small payloads of nearly 45 kg into LEO. The program is planned to drive down liftoff expenses for the United States military small satellites to nearly US300, 000 dollars per launch and if the program of development was funded (Sarnikorpi, 2017).

The Swiss company S3 (Swiss Space Systems) publicized its plans in the year 2013 to develop the suborbital spaceplane termed as *SOAR* that would liftoff the microsat launch vehicle having the capability of putting the payload of nearly 250 kg into LEO.

7.2.5. Nanosatellites

The term "nanosat" or "nanosatellite" generally refers to the artificial satellite having a wet mass between 1 and 10 kg. Proposed designs and designs of these kinds might be launched separately, or they might have various nanosats functioning together or in a formation, in which situation, at times the term satellite swarm or fractionated spacecraft might be applied (Long, 2005). Some of the designs need a larger mother satellite in order to communicate with the ground controllers or for putting into orbit and docking with the nanosatellites. Around 1300 nanosatellites have been put into orbit till January 2021 (Figure 7.3).

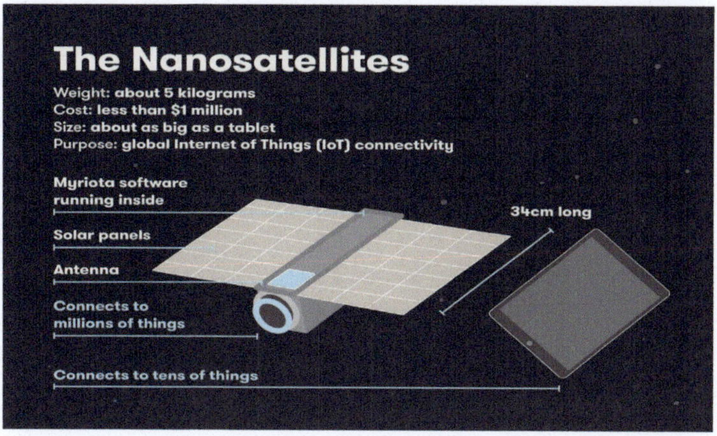

Figure 7.3. *Characteristics of the nanosatellite.*

Source: https://www.prnewswire.com/news-releases/myriota-partners-with-tyvak-to-develop-and-launch-next-generation-nanosatellites-300790466.html.

A CubeSat is the common type of nanosat, developed in the form of a cube centered on manifolds of 10 cm × 10 cm × 10 cm, having a mass of nearly 1.33 kilograms per unit. The concept of CubeSat was first established in 1999 by the joint team of Stanford University and California Polytechnic State University, and specifications, for utilization by anyone intending to launch the CubeSat-style nanosat, are preserved by this group (Knight *et al.,* 2001).

The wet mass of these satellites is the range of 1–10 kilograms. They are launched as an individual as well as clusters of satellites developed to work together in order to perform the anticipated tasks. Some of the designs need the larger satellite, usually known as the mother satellite, for communication with the ground controllers and for putting into orbit and docking with the nanosatellites. The term satellite swarm is usually used to designate the group of nanosatellites. With developments in electronics technology, chiefly in size reduction for the given operational capability, the nanosatellites are quickly taking the space for uses that previously needed the utilization of mini and microsatellites (Pinciroli et al., 2008; King *et al.,* 2012). One such instance is that of the constellation of nearly 35 nanosatellites each one of them weighing 8 kg replacing the constellation of 5 Rapid-Eye satellites each one of them weighing around 156 kg for the Earth-imaging uses at the similar mission cost and considerably increased the revisit time of around 3.5 hours as against the 24 hours. Some of the other instances of the nanosatellite missions comprise BRITE and UniBRITE, which make part of the constellation of 6 BRITE satellites, developed to make accurate measurements of brightness variations of a large number of very bright stars, AAUSAT-2 and AAUSAT-3, designed by the Aalborg University of Denmark, and STRAND, designed by the University of Surrey and the ELaNa mission of the NASA incorporation with diverse universities to liftoff small satellites for the research purposes (Butrica, 2006; Storck *et al.*, 2006).

A term generally used with pico and nanosatellites is CubeSat. CubeSat is the kind of small satellite of nearly 10 centimeters cube dimensions and a mass that is not greater than 1.33 kilograms. The standard 10 × 10 × 10 centimeters basic CubeSat is known as the 1U CubeSat, where this 1U stands for 1 unit (Figure 7.4).

Figure 7.4. *ELaNa.*

Source: Image courtesy: NASA; http://www.spaceref.com/news/viewpr.html?pid=35029.

CubeSats are generally scalable in additions of 1U along 1 axis only. Therefore, 4U, 3U, and 2U CubeSats will generally have the dimensions of 40 × 10 × 10 centimeters, 30 × 10 × 10 centimeters, and 20 × 10 × 10 centimeters, correspondingly. Since all of the CubeSats are normally 10 × 10 cm regardless of length, all of them can be put into orbit and deployed from the common deployment system (Akhtar and Linshu, 2006; Hajiyev and Soken, 2014).

With continuous progress in the smallness and capability upsurge of the electronic technology and utilization of the satellite constellations, nanosats are increasingly proficient in executing commercial missions that earlier needed microsatellites. For instance, the standard of 6U CubeSat has been offered to enable the constellation of 35 satellites having 8 kg Earth-imaging satellites to substitute the constellation of 5,156 kg. Rapid-Eye Earth-imaging satellites, at similar mission price, with considerably augmented revisit times: every portion of the world can be pictured every 3.5 hours instead of once per day with RapidEye constellation. Quicker revisit times are a significant enhancement for nations carrying out disaster response, which was the aim of the Rapid-Eye constellation. Furthermore, the nanosatellite option would permit more countries to have their satellite for off-peak imaging data gathering. With lower costs and shorter production times, nanosats are becoming progressively possible ventures for companies (Babuscia, 2020). Examples of nanosatellites include ExoCube (CP-10), SPROUT, and ArduSat. Nanosatellite manufacturers and developers include NanoAvionics, GomSpace, NanoSpace, Surrey Satellite Technology, Spire, Nova Wurks, Planet Labs, Dauria Aerospace, and Reaktor (Pelton, 2019).

7.2.5.1. Nanosat Market

In the 10 years of nanosatellite launches before 2014, just 75 nanosats were put into orbit. The rates of Launch picked up considerably when in the 3-month era from November 2013 to January 2014, 94 nanosatellites were launched. One main challenge of utilizing nanosatellites has been the cost-effective delivery of these small satellites to anyplace beyond LEO. By late 2014, the proposals were being made for larger spacecraft specially developed to transport swarms of nanosatellites to trajectories that are far beyond Earth orbit for uses like exploring distant asteroids (Weinzierl, 2018).

7.2.5.2. Nanosat Launch Vehicle

With the appearance of technological progress of miniaturization and augmented capital to fund private spaceflight creativities in the 2010s, various startups have been made to pursue chances with developing the range of small-payload NLV (nanosatellite launch vehicle) technologies.

NLVs under development or proposed include (Mosier, 1992):

i. Ventions' Nanosat upper stage.

ii. The upper stage of Virgin Orbit *LauncherOne*, meant to be air-launched from the White Knight 2 comparable to how the space plane SpaceShipTwo is launched.

iii. Andøya/Nammo *North Star*

iv. Till April 2013, the Garvey Spacecraft (now termed as Vector Launch) is developing their *Prospector 18* sub-orbital launch vehicle technology into the orbital NLV having the capability of delivering the 10 kg payload into the 250 km orbit.

v. Generation Orbit is emerging an air-launched rocket in order to transport both sub-50 kg microsats and nanosats to LEO.

Authentic NS launches (Gruss, 2015):

- NASA launched 3 satellites in April 2013 centered on smartphones. Two phones utilize the specification of PhoneSat 1.0 and the 3rd used the beta version of PhoneSat 2.0.
- ISRO put into orbit 14 nanosatellites in June 2016, 12 for the US under the program of Flock-2P and two for the Indian universities. This launch was carried out during the PSLV-C34 mission.
- ISRO put into orbit 103 nanosatellites in February 2017. This liftoff was carried out during the PSLV-C37 mission.

7.2.6. Picosatellites

The term "picosat" or "picosatellite" usually refers to the artificial satellites having a wet mass between 0.1 and 1 kg, even though it is at times utilized to refer to a satellite that is less than one kilogram in launch mass. Proposed designs and designs of these kinds usually have several picosats functioning together or in a formation (at times, the term swarm is used). Some designs need a larger mother satellite in order to communicate with the ground controllers or for putting into orbit and docking with the picosatellites (Figure 7.5) (Shimada et al., 2014).

Figure 7.5. *The three-dimensional model of the picosatellite.*

Source: https://researchtrustmalta.eu/tag/picosatellites/.

Picosatellites are evolving as a novel substitute for doing it yourself, kit builders. The opportunities of a launch are now obtainable for $12,000–18,000 for sub-one kilogram picosatellite payloads that are nearly the size of the soda can.

The wet mass of these satellites is the range of 0.1–1 kilogram. They are normally launched as a group of satellites to operate in formation for anticipated mission aims. As for the nanosatellites, some of the designs of the picosatellites also need the larger mother satellite generally for communication with the ground controllers and for putting into orbit and docking with the picosatellites. The typical design of CubeSat described in the prior section is an instance of the smallest nanosatellite or the largest picosatellite (Schilling, 2006; Tubbal *et al.,* 2015).

Pico satellites provide an outstanding way for PhD and graduate students in order to get experience with satellite system design. One instance is the satellite UWE-1 (experimental satellite of the University of Wurzburg) with a mass of generally less than one kg. The satellite was designed and built by the students. It was put into orbit in the year 2005 in order to test versions of the internet protocols (IPs) to the environment of space, characterized by substantial signal propagation delays because of the large distances and higher noise levels than the terrestrial links (Ortega et al., 2010; Dumont *et al.,* 2013).

7.2.7. Femtosatellites

The term "femtosat" or "Femto satellite" generally refers to the artificial satellites having a wet mass of less than 100 g. Like picosatellites, some of the designs need a larger mother satellite in order to communicate with the ground controllers.

Three prototype chipsats were launched to International Space Station (ISS) on the Space Shuttle *Endeavour* in May 2011 on its last mission. They were fixed to the external platform MISSE-8 (Materials International Space Station Experiment) of ISS for testing. In April 2014, the nanosat KickSat was put into orbit aboard the Falcon 9 rocket having the intention of freeing 104 femtosat-sized chipsats, or the Sprites. In the occurrence, they were incapable to finish the deployment on due time because of the failure of the aboard clock and the mechanism of deployment entered the atmosphere again on 14 May 2014, without deploying any of the five-gram femtosats. ThumbSat is one more project which launched femtosats in the late 2010s (Figure 7.6) (Rugescu et al., 2014).

Figure 7.6. *Next-generation of sensors for the femto-satellites.*

Source: https://www.researchgate.net/figure/NextGen-of-sensors-for-femto-satellites_fig1_254009029.

In March 2019, the CubeSatellite KickSat-2 deployed 105 femtosatellites known as "ChipSats" into the Earth orbit. These satellites were tested for three days and they then the atmosphere again and burned up.

The wet mass of these satellites lies in the range of 10–100 grams. Again, like pico and nanosatellites, some of the designs need a larger mother satellite in order to communicate with the ground controllers. With progress in nano and microtechnologies, today it is possible to make satellite subsystems and the complete satellite on a chip. Such kind of satellites is called chipsats (Houborg and McCabe, 2016).

Three prototype chipsats were launched to ISS on the Space Shuttle Endeavor in May 2011 on its last mission. The small chip satellites, named Sprite, were mounted on a MISSE-8 pallet, which was fixed to the ISS to test how better they functioned in the harsh environment.

The satellites having the size of fingernails were developed to gather the chemistry of solar wind, radiation, and particle-impact data. In one more instance, the KickSat mission aims to launch 250 small Sprite satellites into the LEO. The launch is intended on the launch vehicle Falcon-9 of SpaceX in early 2014. This mission will give a chance to the individuals to have their satellites at the price of around US 300 dollars per satellite (Figure 7.7) (Sweeting, 2000; Kim *et al*., 2010).

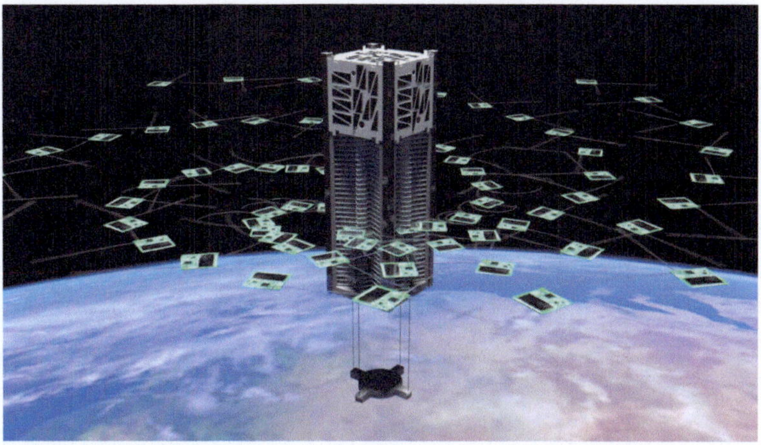

Figure 7.7. *KickSat mission.*

Source: https://kicksat.github.io/.

7.2.8. Technical Challenges

Small satellites generally need attitude control, innovative propulsion, computation, and communication systems.

Larger satellites normally utilize bipropellant or monopropellant combustion systems for attitude control and propulsion; these systems are intricate and need a negligible amount of volume-to-surface area in order to disperse heat. These systems might be utilized on larger small satellites, whereas other nano/microsats have to utilize electric propulsion, vaporizable liquids, compressed gas, like carbon dioxide or butane, or other inventive systems of propulsions that are cheap, simple, and scalable (Jennewein *et al.*, 2014).

Small satellites can utilize traditional radio systems in VHF, UHF, X-band, and S-band, even though often miniaturized utilizing more modern technology compared to the larger satellites. Miniaturized satellites like nanosats and small microsats might lack the mass or power supply for the large conventional radio transponders and several miniaturized or advanced systems of communication have been suggested, like antenna arrays, laser receivers, and satellite-to-satellite networks of communication (Handberg, 2014).

Electronics must be strictly tested and altered to be space hardened or resilient to the environment of outer space. Miniaturized satellites permit the chance to test novel hardware with decreased cost in testing. Moreover, as the overall expense risk is much lower in the mission, modern but less space-proven expertise can be integrated into nano and microsats than can be utilized in larger, more costly missions with less desire for risk.

7.3. TEN-KOH PROJECT

The small satellite mission named The-Koh had the following main objectives (Martínez and León, 2018; Piskorz and Jones, 2018):

- To illustrate the environment of plasma around the spinning spacecraft;
- To identify MeV-range electrons in Leo Earth orbit and examine the environment of space in the existence of the low solar activity;
- To examine the variation of the physical properties of CFRP and LATS material samples uncovered to the environment of space.

Including students in the development, testing, manufacturing, and functions of a satellite is a significant part of their program and improves their education. Giving this participation has been set as the secondary mission aim of Ten-Koh.

Ten-Koh also offered the flight opportunity for two technology demo payloads, which established another secondary aim of the mission. The payloads are Thermal Switch designed at S3, which flight-proves a new design of the switchable thermal switch, and an Ultra-capacitor Experiment that intends to measure the performance of the ultra-capacitor as the satellite energy storage device (Kulu, 2019).

7.3.1. Components

Figures 7.8–7.10 display the Ten-Koh satellite in the in-orbit configuration. The platform of a satellite is made of the succeeding subsystems (Verhoeven et al., 2011; Zarifian et al., 2015):

1. **COMM:** Communication subsystem;
2. **OBC:** Onboard computer and data handling subsystem;
3. **ADS:** Attitude determination subsystem;
4. **EPS:** Electrical power subsystem.

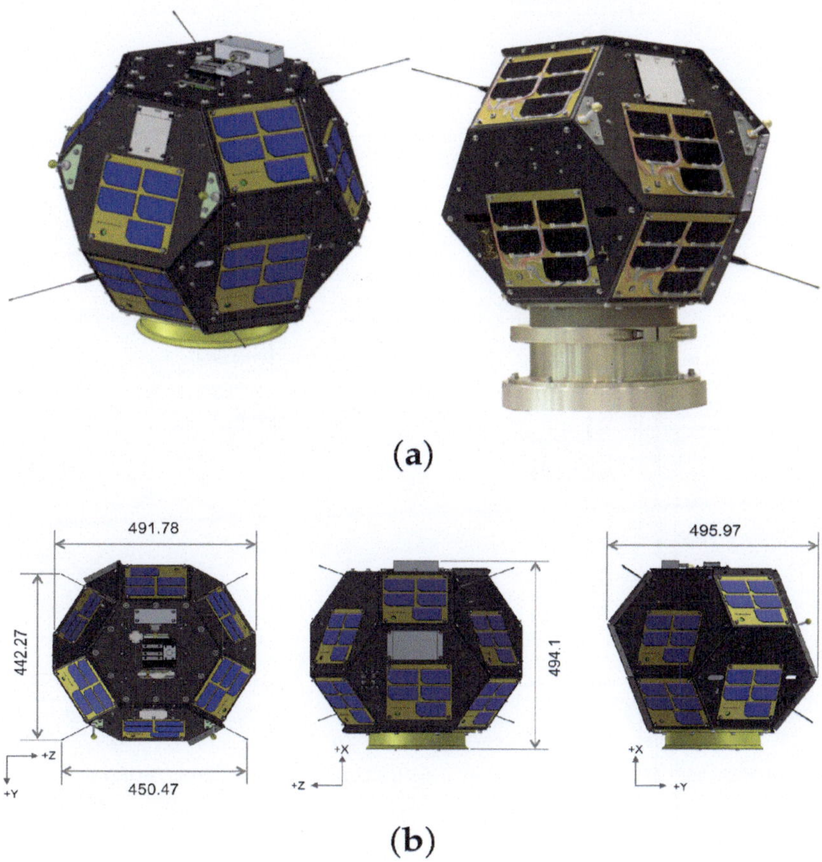

Figure 7.8. *Flight model configuration of Ten-Koh satellite in orbit. The black outer structure is formed of CFRP. (a) On the left is the computer-aided design and on the right is the flight model photograph; (b) the envelope permitted by the launch vehicle was 500 ×500 × 500 mm.*

Source: https://www.researchgate.net/publication/336111166_Design_Implementation_and_Operation_of_a_Small_Satellite_Mission_to_Explore_the_Space_Weather_Effects_in_Leo.

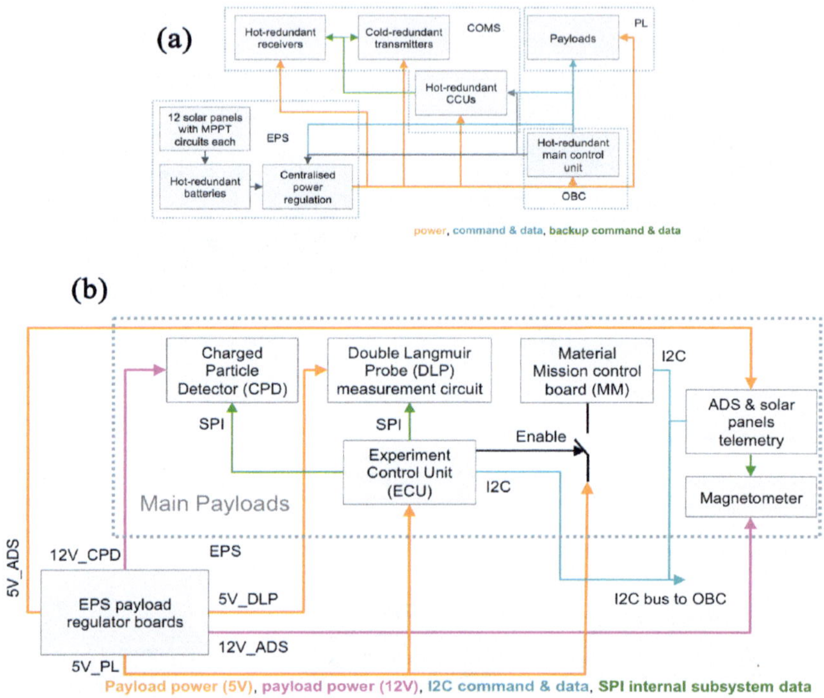

Figure 7.9. *(a) System architecture of Ten-Koh system; (b) Ten-Koh payload architecture comprising: ECU (experiment control unit), MM (material mission), CPD (charged particle detector), and DLP (double Langmuir probe).*

Source: https://www.semanticscholar.org/paper/Design%2C-Implementation%2C-and-Operation-of-a-Small-to-Fajardo-Lidtke/7137d95d027c0c20caec24e37e1753c139d116f2.

Note: Block diagrams of the Ten-Koh system along with its payloads. The dashed lines exhibit the limit of every subsystem, and the type of interface is indicated by the solid lines. Ten-Koh satellite utilizes I2C as the key data bus (blue), and the backup data bus is SPI (gray) and direct interface with the sensors inside each subsystem. The 5 Volts power lines are displayed in yellow color and the 12 Volts lines in magenta. The subsystem of ADS was comprised inside the payload (PL) subsystems for suitability in a physical location within the satellite, therefore it looks in the block diagram at the bottom.

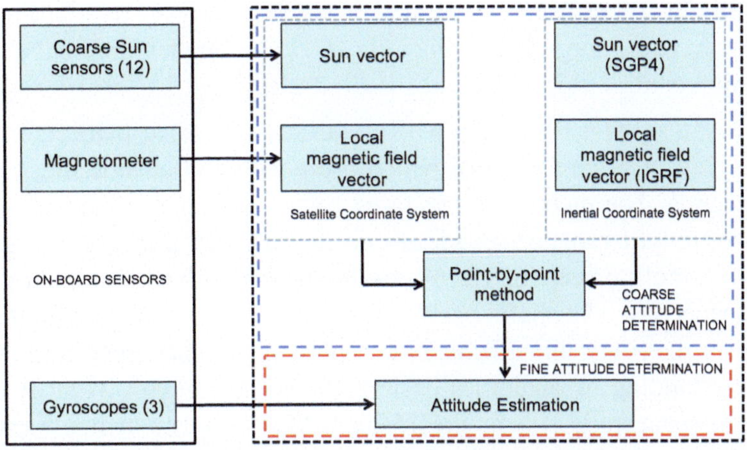

Figure 7.10. *Philosophy of attitude determination for a Ten-Koh satellite. The magnetometer is utilized for space environment purposes along with ADS.*

Source: https://www.mdpi.com/2226-4310/6/10/108/htm.

7.3.1 is not necessary

7.3.2. Payload

The key payload comprises five individual instruments. One main payload is the CPD (charged particle detector), which was made by TX, Prairie View A&M University, United States, and the Institute of Space Research and Technology, Bulgaria. The system comprises of eight complementary metal-oxide semiconductor (CMOS) detectors mounted on five faces of the cube, and a Bulgarian Liulin-kind detector fixed on the top of the entire assembly. The CPD instrument backs to all four main mission objectives (Swartwout, 2016; Logue and Pelton, 2019).

One more payload of the Ten-Koh mission is the set of DLP (Double Langmuir Probes). This experiment is meant to collect information regarding plasma and the sheath creation around a satellite as it is rotating along its orbit. The charging of satellite, which can take place at the internal level and the surface, is a significant subject of study since it is linked to the anomalies of satellite and is directly affected by the weather conditions of space. Satellite charging is dependent on the collaboration of spacecraft materials, thickness of materials, and the energy of charged particles. Fluxes of electrons varying from 10–100 keV can yield surface charging, whereas electrons having energy bigger than the 100 keV can yield internal charging (McKnight et al., 2015; Chen and Li, 2018).

The system of DLP is made up of 2 spherical, 10 μm Au-plated electrodes, as displayed in Figure 7.11(a), which are positioned outside the structure of spacecraft. The electrodes are fixed on an isolating GFRP (glass fiber reinforced polymer) plate to evade any electrical interaction with the structure of CFRP, therefore keep the electrodes fluctuating with respect to the ground of the spacecraft. The shafts backing the spheres are covered with alumina to confirm that only the spheres entice plasma particles, therefore making the understanding of outcomes easier. The control circuit of DLP, placed inside the spacecraft, utilizes the 16-bit DAC (digital-to-analog converter) to produce the DC biasing voltage. This permits the duration of sweep to be altered such that plasma in diverse locations around the revolving spacecraft can be categorized. The current flow amongst the probes over the plasma and biasing voltage are then measured with the AD7927 12-bit ADC. The circuit of measurement is separated from the spacecraft's common ground, which permits the parameters of plasma to be measured.

The objective of the MM (material mission) is to uncover three different samples of a novel material for the space applications made of PEEK resin and carbon fiber and developed by the Okuyama laboratory of Kyutech (Tsitas and Kingston, 2012):

- • **Sample 1:** PEEK/CF with no coating.
- • **Sample 2:** PEEK/CF with the special coating silsesquioxane to safeguard against atomic oxygen.
- • **Sample 3:** PEEK/CF with the special coating yttrium oxide to safeguard against UV.

The system measures diverse parameters, comprising temperature and the strain inside samples, to quantify the variations in the CTE (coefficient of thermal expansion) of the novel materials. By measuring the variations in CTE, the internal micromechanical alterations in the chemical composition of samples instigated by the effects of the environment of space can be distinguished. Testing in this manner permits the assessment of material reliability and survivability for utilization in a space environment. The temperatures are measured with AD590 transducers sampled by an AD7927 ADC (Figure 7.12) (Länsiluoto et al., 2004; Pereira et al., 2020).

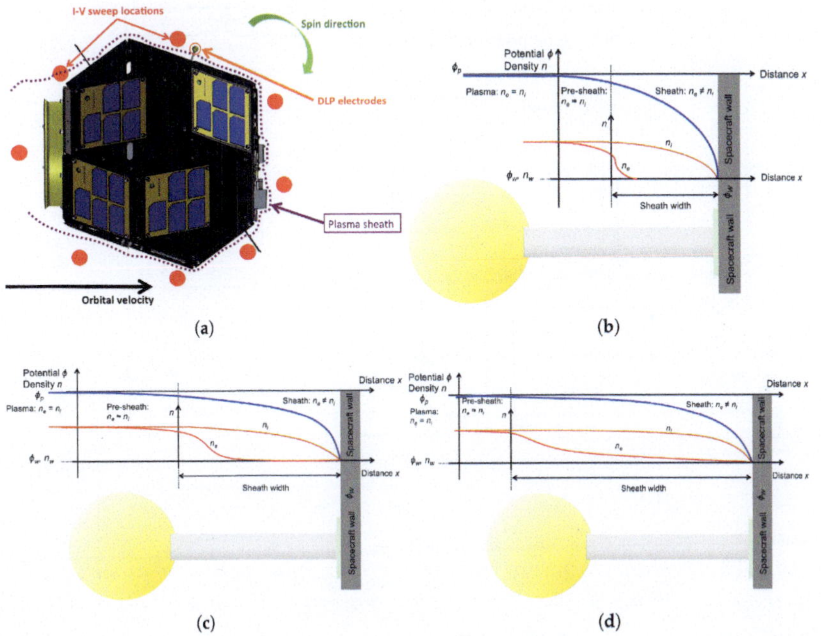

Figure 7.11. *Demonstration of the double Langmuir probes measuring system: (a) current-voltage sweep locations around the revolving spacecraft; and (b to d) one-dimensional model demonstrating the electrodes outside and inside of the plasma sheath.*

Source: https://www.mdpi.com/2226-4310/6/10/108/htm.

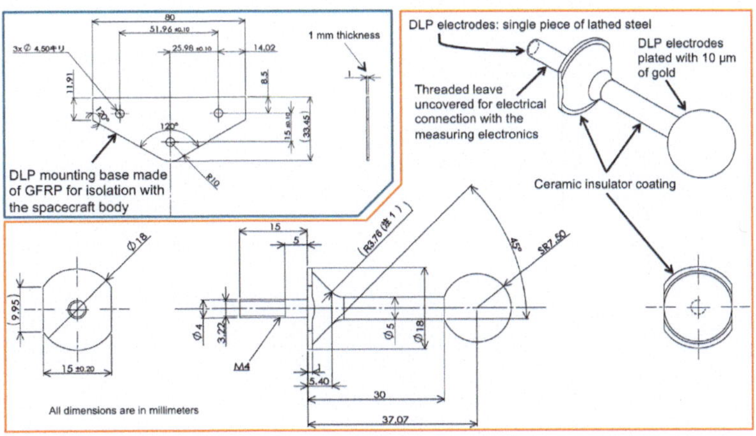

(a)

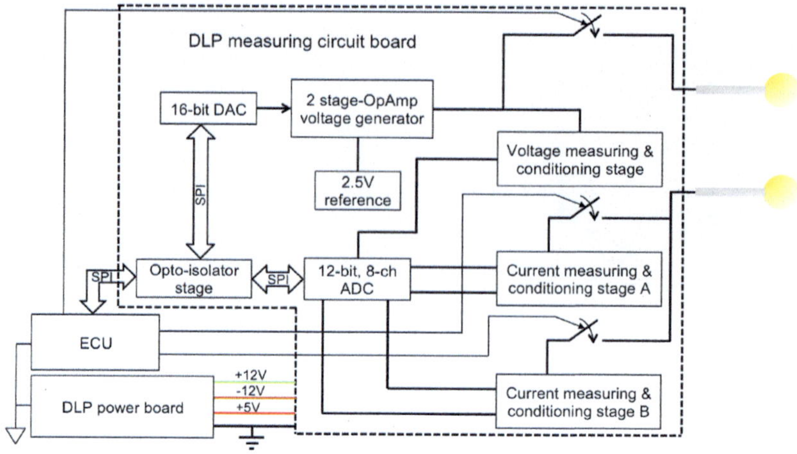

(b)

Figure 7.12. *Demonstration of the double Langmuir probes measuring system: (a) double Langmuir probes electrode composed of solid aluminum, covered with 10 µm of Au in a spherical tip, the ceramic material in a shaft for separation, and FR4 plate on base for isolation purposes from the spacecraft's structure; and (b) the block diagram of double Langmuir probes measuring circuit. The dashed line specifies the double Langmuir probes measuring circuit physically employed in the single PCB.*

Source: https://www.researchgate.net/publication/336111166_Design_Implementation_and_Operation_of_a_Small_Satellite_Mission_to_Explore_the_Space_Weather_Effects_in_Leo.

REFERENCES

1. Akhtar, S., & Linshu, H., (2006). An efficient evolutionary multi-objective approach for robust design of multi-stage space launch vehicle. In: *11th AIAA/ISSMO Multidisciplinary Analysis and Optimization Conference* (Vol. 11, pp. 7073).
2. Angelopoulos, V., Runov, A., Zhou, X. Z., Turner, D. L., Kiehas, S. A., Li, S. S., & Shinohara, I., (2013). Electromagnetic energy conversion at reconnection fronts. *Science, 341*(6153), 1478–1482.
3. Antonello, R., Branz, F., Sansone, F., Cenedese, A., & Francesconi, A., (2020). High-precision dual-stage pointing mechanism for miniature satellite laser communication terminals. *IEEE Transactions on Industrial Electronics, 68*(1), 776–785.
4. Babuscia, A., (2020). Telecommunication systems for small satellites operating at high frequencies: A review. *Information, 11*(5), 258.
5. Bonnici, M., Mollicone, P., Fenech, M., & Azzopardi, M. A., (2019). Analytical and numerical models for thermal related design of a new pico-satellite. *Applied Thermal Engineering, 159*, 113908.
6. Bouwmeester, J., & Guo, J., (2010). Survey of worldwide pico-and nanosatellite missions, distributions, and subsystem technology. *Acta Astronautica, 67*(7, 8), 854–862.
7. Buchen, E., & DePasquale, D., (2014). *Nano/Microsatellite Market Assessment* (Vol. 1, pp. 1–33). Space Works Enterprises. Inc. (SEI), Atlanta, GA.
8. Butrica, A. J., (2006). Reusable launch vehicles or expendable launch vehicles? a perennial debate. *Critical Issues in the History of Spaceflight, 1*, 1–33.
9. Chen, B. R., & Li, S., (2018). Prehire screening and subjective performance evaluations. *Management Science, 64*(10), 4953–4965.
10. Chen, L., Thorne, R. M., Li, W., Bortnik, J., Turner, D., & Angelopoulos, V., (2012). Modulation of plasmaspheric hiss intensity by thermal plasma density structure. *Geophysical Research Letters, 39*(14).
11. Cucinotta, F. A., Cacao, E., Kim, M. H. Y., & Saganti, P. B., (2020). Cancer and circulatory disease risks for a human mission to mars: Private mission considerations. *Acta Astronautica, 166*, 529–536.
12. Dandumont, C., Defrère, D., Kammerer, J., Absil, O., Quanz, S. P., & Loicq, J., (2020). Exoplanet detection yield of a space-based Bracewell

interferometer from small to medium satellites. *Journal of Astronomical Telescopes, Instruments, and Systems, 6*(3), 035004.

13. Dumont, E., Burkhardt, H., Sippel, M., Johannsson, M., Ludwig, C., & David, E., (2013). *Exploiting Technological Synergies for Future Launch Vehicles, 1*, 1–33.
14. Foust, J., (2015*). SmallSat Developers Enjoy Growth in Launch Options* (Vol. 1, pp. 1–29). Space News.
15. Graydon, M., & Parks, L., (2020). 'Connecting the unconnected': A critical assessment of US satellite internet services. *Media, Culture and Society, 42*(2), 260–276.
16. Gruss, M., (2015). *DARPA Scraps Plan to Launch Small Sats from F-15 Fighter Jet* (Vol. 1, pp. 1–22). Space News.
17. Hajiyev, C., & Soken, H. E., (2014). Robust adaptive unscented Kalman filter for attitude estimation of Pico satellites. *International Journal of Adaptive Control and Signal Processing, 28*(2), 107–120.
18. Handberg, R., (2014). Building the new economy: "New space" and state spaceports. *Technology in Society, 39*, 117–128.
19. Houborg, R., & McCabe, M. F., (2016). High-resolution NDVI from planet's constellation of earth-observing nanosatellites: A new data source for precision agriculture. *Remote Sensing, 8*(9), 768.
20. Hussmann, H., Sohl, F., & Spohn, T., (2006). Subsurface oceans and deep interiors of medium-sized outer planet satellites and large trans-Neptunian objects. *Icarus, 185*(1), 258–273.
21. Izquierdo, L., & Tristancho, J., (2011). Next-generation of sensors for Femto-satellites based on commercial-of-the-shelf. In: *2011 IEEE/AIAA 30th Digital Avionics Systems Conference* (Vol. 1, p. 8A4-1). IEEE.
22. Jennewein, T., Grant, C., Choi, E., Pugh, C., Holloway, C., Bourgoin, J. P., & Zee, R., (2014). The nano QEY mission: Ground to space quantum key and entanglement distribution using a nanosatellite. In: *Emerging Technologies in Security and Defense II; and Quantum-Physics-Based Information Security III* (Vol. 9254, p. 925402). International Society for Optics and Photonics.
23. Jones, N., (2014). Mini satellites prove their scientific power. *Nature News, 508*(7496), 300.
24. Kil, H., & Paxton, L. J., (2017). Global distribution of nighttime medium-scale traveling ionospheric disturbances seen by swarm satellites. *Geophysical Research Letters, 44*(18), 9176–9182.

25. Kim, J. S., Jung, H., Kam, H. D., Seo, H. S., & Su, H., (2010). A development of the thrusters for space-vehicle maneuver/ACS and their application to launch vehicles. *Journal of the Korean Society of Propulsion Engineers, 14*(6), 103–120.
26. King, J., Ness, J., Bonin, G., Brett, M., & Faber, D., (2012). *Nanosat Ka-Band Communications-A Paradigm Shift in Small Satellite Data Throughput, 1,* 1–33.
27. Kirby, K., Bushman, S., Butler, M., Conde, R., Fretz, K., Herrmann, C., & Williams, B., (2012). Radiation belt storm probe spacecraft and impact of environment on spacecraft design. In: *2012 IEEE Aerospace Conference* (Vol. 1, pp. 1–20). IEEE.
28. Knight, S., Rabideau, G., Chien, S., Engelhardt, B., & Sherwood, R., (2001). Casper: Space exploration through continuous planning. *IEEE Intelligent Systems, 16*(5), 70–75.
29. Konecny, G., (2004). Small satellites-a tool for earth observation? In: *XXth ISPRS Congress, Commission* (Vol. 4, pp. 12–23).
30. Kulu, E., (2019). Nanosatellite and CubeSat database. *Nanosatellite and CubeSat Database* (Vol. 1, pp. 1–22).
31. Länsiluoto, A., Eklund, T., Back, B., Vanharanta, H., & Visa, A., (2004). Industry-specific cycles and companies' financial performance comparison using self-organizing maps. *Benchmarking: An International Journal, 1*, 1–33.
32. Laštovička-Medin, G., (2016). Nano/Pico/Femto-satellites: Review of challenges in space education and science integration towards disruptive technology. In: *2016 5th Mediterranean Conference on Embedded Computing (MECO)* (Vol. 1, pp. 357–362). IEEE.
33. Logue, T. J., & Pelton, J., (2019). Overview of commercial small satellite systems in the "new space" age. *Handbook of Small Satellites: Technology, Design, Manufacture, Applications, Economics and Regulation, 1*, 1–18.
34. Long, A. M., (2005). *Framework for Evaluating Customer Value and the Feasibility of Servicing Architectures for on-Orbit Satellite Servicing* (Vol. 1, pp. 1–33). Doctoral dissertation, Massachusetts Institute of Technology.
35. Martínez, J. G. D., & León, R. Á., (2018). Bioethical aspects related to space debris and its effects on life on earth and aerospace exploration. *Person and Bioethics, 22*(1), 39–55.

36. McKnight, B. R., Boyer, J. E., Nardozzo, P. K., & Cortopassi, A., (2015). Design and testing of an additively manufactured advanced hybrid rocket motor propulsion unit for CubeSats (PUC). In: *51st AIAA/SAE/ASEE Joint Propulsion Conference* (Vol. 1, p. 4036).
37. Messier, D., (2015). Euroconsult sees large market for SmallSats. *Parabolic Arc* (Vol. 1, pp. 1–33).
38. Miyake, S., Kataoka, R., & Sato, T., (2017). Cosmic ray modulation and radiation dose of aircrews during the solar cycle 24/25. *Space Weather, 15*(4), 589–605.
39. Mosier, M., (1992). Pegasus dedicated and multiple launch capabilities for low earth orbit communications satellite constellations. In: *14th International Communication Satellite Systems Conference and Exhibit* (Vol. 1, p. 1878).
40. Nakamura, Y., Fukuda, S., Shibano, Y., Ogawa, H., Sakai, S. I., Shimizu, S., & Shinohara, I., (2018). Exploration of energization and radiation in geospace (ERG): Challenges, development, and operation of satellite systems. *Earth, Planets and Space, 70*(1), 1–20.
41. Ortega, P., López-Rodríguez, G., Ricart, J., Domínguez, M., Castañer, L. M., Quero, J. M., & Angulo, M., (2010). A miniaturized two axis sun sensor for attitude control of nano-satellites. *IEEE Sensors Journal, 10*(10), 1623–1632.
42. Paziewski, J., Sieradzki, R., & Wielgosz, P., (2018). On the applicability of Galileo FOC satellites with incorrect highly eccentric orbits: An evaluation of instantaneous medium-range positioning. *Remote Sensing, 10*(2), 208.
43. Pelton, J. N., (2019). Trends in chemical rocket systems and new approaches to launching satellites. In: *Space 2.0* (Vol. 1, pp. 129–143). Springer, Cham.
44. Pereira, A., Biddington, B., Suzuki, K., Rajagopalan, R. P., Krishna, R., & Vaccaro, D., (2020). America back to the moon and on to mars: Australian, Japanese, and Indian perspectives. In: *2020 IEEE Aerospace Conference* (Vol. 1, pp. 1–14). IEEE.
45. Pinciroli, C., Birattari, M., Tuci, E., Dorigo, M., Del, R. Z. M., Vinko, T., & Izzo, D., (2008). Self-organizing and scalable shape formation for a swarm of Pico satellites. In: *2008 NASA/ESA Conference on Adaptive Hardware and Systems* (Vol. 1, pp. 57–61). IEEE.

46. Piskorz, D., & Jones, K. L., (2018). On-orbit assembly of space assets: A path to affordable and adaptable space infrastructure. *The Aerospace Corporation* (Vol. 1, pp. 1–28).
47. Richie, D. J., Lappas, V. J., & Palmer, P. L., (2007). Sizing/optimization of a small satellite energy storage and attitude control system. *Journal of Spacecraft and Rockets, 44*(4), 940–952.
48. Rugescu, R. D., Rugescu, D. R., & Micu, E., (2014). Orbital launcher NERVA as the first proof of the discontinuous variational solution for the atmospheric ascent. In: *Applied Mechanics and Materials* (Vol. 555, pp. 91–101). Trans Tech Publications Ltd.
49. Sarnikorpi, J., (2017). *Working Environments in NewSpace Product Development: A Case Study of a Nanosatellite Startup, 1*, 1–33.
50. Schilling, K., (2006). Design of pico-satellites for education in systems engineering. *IEEE Aerospace and Electronic Systems Magazine,* 21(7), S_9–S_14.
51. Sharma, A. S., & Curtis, S. A., (2005). Magnetospheric multiscale mission. In: *Nonequilibrium Phenomena in Plasmas* (Vol. 1, pp. 179–195). Springer, Dordrecht.
52. Shimada, T., Kitagawa, K., Kanazaki, M., Yuasa, S., Sakurai, T., Nakagawa, I., & Ozawa, K., (2014). Study of low-cost 100 kg-satellite launcher using hybrid rocket engines. *Session 7 of Space Propulsion, 1*, 1–33.
53. Sihver, L., Kodaira, S., Ambrožová, I., Uchihori, Y., & Shurshakov, V., (2016). Radiation environment onboard spacecraft at LEO and in deep space. In: *2016 IEEE Aerospace Conference* (Vol. 1, pp. 1–9). IEEE.
54. Somov, Y., Sukhanov, V., & Hacizade, C., (2014). Guidance and precise motion control of free-flying robots and land-survey mini-satellites. In: *2014 13th International Conference on Control Automation Robotics and Vision (ICARCV)* (Vol. 1, pp. 1092–1097). IEEE.
55. Spector, S., (2020). The cosmic-local nexus: Intersections between outer space and rural communities. *Tourism Recreation Research, 45*(1), 94–106.
56. Storck, W., Billett, O., Jambusaria, M., Sadhwani, A., Jammes, P., & Cutler, J., (2006). *A Survey of Micropropulsion for Small Satellites, 1,* 1–44.
57. Stratton, J. M., Harvey, R. J., & Heyler, G. A., (2013). Mission overview for the radiation belt storm probes mission. *Space Science Reviews, 179*(1–4), 29–57.

58. Suparta, W., (2014). The variability of space radiation hazards towards LEO spacecraft. In: *Journal of Physics: Conference Series* (Vol. 539, No. 1, p. 012023). IOP Publishing.
59. Swartwout, M., (2016). Assessing the debris risk from the avalanche of CubeSats. In: *2016 IEEE Aerospace Conference* (Vol. 1, pp. 1–9). IEEE.
60. Sweeting, M. N., (2000). Space at surrey: Micro-mini-satellites for affordable access to space. *Air and Space Europe, 2*(1), 38–52.
61. Tristancho, J., & Gutierrez-Cabello, J., (2011). A probe of concept for Femto-satellites based on commercial-of-the-shelf. In: *2011 IEEE/AIAA 30th Digital Avionics Systems Conference* (Vol. 1, pp. 8A2-1). IEEE.
62. Tristancho, M. J., (2010). *Implementation of a Femto-Satellite and a Mini-Launcher for the N Prize* (Vol. 1, pp. 1–33).
63. Tsitas, S. R., & Kingston, J., (2012). 6U CubeSat commercial applications. *The Aeronautical Journal, 116*(1176), 189–198.
64. Tubbal, F. E., Raad, R., & Chin, K. W., (2015). A survey and study of planar antennas for pico-satellites. *IEEE Access, 3*, 2590–2612.
65. Verhoeven, C. J., Bentum, M. J., Monna, G. L. E., Rotteveel, J., & Guo, J., (2011). On the origin of satellite swarms. *Acta Astronautica, 68*(7, 8), 1392–1395.
66. Walker, L., Di Carlo, M., Greco, C., Vasile, M., & Warden, M., (2020). A mission concept for the low-cost large-scale exploration and characterization of near-earth objects. *Advances in Space Research, 1*, 1–43.
67. Weinzierl, M., (2018). Space, the final economic frontier. *Journal of Economic Perspectives, 32*(2), 173–192.
68. Werner, D., (2013). *Small Satellites and Small Launchers; Rocket Builders Scramble to Capture Growing Micro Sat Market* (Vol. 1, pp. 1–3). Space News.
69. Wisnarama, S. I., (2014). *A Large-Scale Simulation of Satellites Tracking Vessels and Other Targets* (Vol. 1, pp. 1–24).
70. Woerd, H. J., & Wernand, M. R., (2015). True color classification of natural waters with medium-spectral resolution satellites: Sea WiFS, MODIS, MERIS and OLCI. *Sensors, 15*(10), 25663–25680.

71. Zarifian, P., Imken, T., Matousek, S. E., Moeller, R. C., Bennett, M. W., Norton, C. D., & Kawata, J., (2015). Team Xc: JPL's collaborative design team for exploring CubeSat, NanoSat, and small sat-based mission concepts. In: *2015 IEEE Aerospace Conference* (Vol. 1, pp. 1–10). IEEE.

Chapter

8

FUTURE TRENDS IN SATELLITE COMMUNICATION SYSTEMS

CONTENT

8.1. Introduction 188

8.2. Current Trends in Satellite Technology 189

8.3. The Path Forward 190

8.4. Satellite Orbital Configurations and Improved Spacecraft Orientation and Pointing 202

8.5. New Ground User Systems 204

8.6. Integrated Satellite and Terrestrial Services and New Market Demand 205

8.7. Tracking, Telemetry, Monitoring, Command, and Autonomous Operations 207

8.8. Future Trends for Markets and Regulatory Systems 208

8.9. New Initiatives in Space Communications 210

8.10. Keys to the Future of Satellite Communications 211

References 216

8.1. INTRODUCTION

Satellite Communications technologies have attained outstanding breakthrough efficiencies and grow in performance in roughly a half-century. However, these advancements have taken place in parallel with a huge boost in performance through other IT and telecommunications systems. Hence, such dramatic gains are not as visible to the general public as might have been the situation if this outbreak in performance had occurred in isolation (Cox and Coney, 1996; O'Dell *et al.*, 2015).

Today's satellites served as digital processors in the sky in various ways, and specialized software specifies their communications abilities and explains how they perform. Similar courses have been followed by the novelty in satellite communications and also the progression in all kinds of computer processes and telecommunications. Briefly, Moore's law that anticipated a doubling of performance after every 18 months has held generally for all fields incorporating digital processing, either it be video games, communications, computing, or even digital entertainment systems. What had been former is therefore assumed to be prolog. It is acceptable to expect chronic gains in terms of digital communications, overall processing power, and 'intelligent' space communication systems (Sherwood et al., 2003; Atzori *et al.*, 2010).

Shortly, there are significant new technologies that are still to be developed in terms of new encoding capabilities, more powerful processors, space-based satellite communications systems, and new user terminal abilities that can form user systems more personally responsive, more versatile, more mobile, more powerful concerning performance, and still less costly (Pelton, 2005; Iida *et al.*, 2003).

Since the world's national economies become too universal and since all segments of the globe, i.e., the atmosphere and the oceans are used by the human enterprise, there will be a requirement of an expansion of effective wireless interconnection through satellite communications and terrestrial wireless. Moreover, the increased usage of space systems-unmanned, manned, and planetary bodies will grow the need for enhanced space communications systems. Undoubtedly, foreseeable technologies recommend that many decades of continuing developments are now possible. But for the satellite communications industry, technology will not be the just source of change. Other drivers of change will involve (Rahmat-Samii and Densmore, 2014; Vasavada *et al.*, 2016):

- New service demands in both defense-related and civilian markets;
- Restructuring of commercial satellite organizations by merger, acquisition, and regulatory change;
- New allocations or reallocation of frequencies;
- Convergence within the different satellite applications markets-both concerning structural and technology integration;
- Constraints in orbital configurations;
- Orbital debris; and
- Increase of human endeavor in outer space may prove to be notable shapers of the increase in satellite systems in the later 20–30 years (Pelton, 2005; Iida et al., 2003).

8.2. CURRENT TRENDS IN SATELLITE TECHNOLOGY

A very powerful gain in performance is represented by today's communications satellites when compared to those initially installed about a half-century ago. Contemporary satellites solar arrays can produce well aloft 100 times more power, and the equivalent of up to a 1,000 times utilizable bandwidth can be given by the state-of-the-art satellite antenna systems as compared to that of the Early Bird Satellite-the globally first commercial satellite spacecraft. Photovoltaic (PV) cells have enhanced in performance, deployable solar arrays have become greater in size, and improvements in design have permitted the arrays to obtain maximum exposure to the sun. Battery systems have also upgraded greater in longevity and power density. Satellite antenna systems have developed and improved in various ways. These have involved multi-beam antennas, better focusing, and pointing of radio frequency (RF) energy, enhanced large-scale antenna manufacturing, and frequency reuse strategies. Overall, the lifetime and performance of satellite systems in space have been increased by an ongoing series of technological advancements and have formed the user equipment on the ground easier to use, lower in cost, and more accessible (Pelton, 2006; Venkatesan *et al.,* 2013).

The outlook implies that numerous of these powerful trends will carry on. However, there are some key challenges. In the previous section, one of these challenges that have been explained is the integration of satellite communications systems with broadband fiber and terrestrial wireless and coaxial cable systems. An alternate challenge is to adapt satellite technology to a dynamic world. This could imply many things. It implies the effective usage of satellite systems to communicate across the Planet and also to direct beyond all over the Solar System (European Space Agency (ESA), 2008). It could mean highly integrated space applications so that user devices could facilitate not only data, video, and voice signals, but also space navigation and location services, weather, and meteorological data, Earth-imaging data, and other required information on demand (Farserotu and Prasad, 2000; Ohata *et al.,* 2005).

In contrast to new satellite technologies, the future will therefore be shaped more by market demands and new services. Certainly, major changes in the satellite communications industry could also be dictated by the constraints imposed by orbital debris, industrial consolidation, regulatory shifts, and even a change in insurance and financial markets (Taleb *et al.,* 2005; Toyoshima, 2005).

Some might advise that today satellite communication systems have transformed from giant 30 m, multi-ton Earth stations to handheld transceiver devices there is slight additional room for further advancements. But numerous times in the past, history has proven forecasters incorrect. Forecasters like Thomas Watson-Chairman of IBM, once believed that hardly a few dozens of computers would be required by the world to be used by the exclusive scientists. Others believed that due to the wind resistance, trains would move no faster than 100 km per hour (or around 60 miles/h). It was advised in the nineteenth century that as all the major inventions had been registered already so the patent offices could be out of service. Demand for new capabilities and new services in a human society constantly provides an increase in novel technologies which following produce new applications and the process reproduces itself repeatedly. Sometimes devotion to technology also overestimates future trends. The technological innovation can be outweighed in obtained correct forecasts for the field of satellite communications projecting demand for newel services. Certainly, forecasts based on technological innovation can sometimes be wrong and also repeatedly greatly emphasized (Buehler and Griffin, 1989).

8.3. THE PATH FORWARD

The field of satellite communications is pretty technical and complicated, but the dynamic range of physical systems within which the contemporary satellite networks are described is so small. Antenna systems spotlight "radio frequencies" and "power." To increase the performance of the antennas, one must concentrate power more effectively or have an approach to more power or determine a way to use available frequencies more efficiently, either by more effective "reuse of the frequencies" or in higher frequency ranges or by both. These are the variety of tools available to form satellite communications more productive. Of course, more effective means can be invented to transmit more "usable information" through a communications channel, either that be a terrestrial wireless link, a fiber optic link, or a satellite. The way forward basically lies along with one of these routes. Hence, this section probes the future in terms of improved power, more effective antennas, more effective spacecraft design-involving improved lifetime, pointing systems, reliability, improved transmission capabilities, satellite orbital configurations, and improved user transceivers (Toyoshima, 2005; Voigt *et al.*, 2016).

8.3.1. Advanced Spacecraft Antenna Design

The key to a satellite antenna's performance includes how well it can meet an RF beam approaching the "catchment area" or designed reception. This property of the antenna is known as antenna gain. A narrower beam can be created from a larger aperture antenna, and therefore there is slight path loss because of the signal spreading within the Earth and satellite. For the creation of the various hundreds of narrow and highly focused beams, currently, the largest aperture satellite antennas having diameters on the order of 20 m or more can be utilized in combination with a multi-feed system, which permits intensive frequency reuse (Ruiz-Garcia *et al.*, 2009; Emrick *et al.*, 2014). This kind of progressive multi-beam, large aperture communications satellite antenna can be noticed in such spacecraft as the Skyterra, TerreStar, and Inmarsat 4 satellites. This is because the same frequencies can be used repeatedly by the beams that are geographically detached from one another. Naturally, the question comes to mind as to just to which extent can satellite communications space antennas evolve without their being structural or cost barriers to their forthcoming expansion? (Pelton, 1998; Iida and Suzuki, 2001).

The answer to this question shows up rather complex in that there is a range of approaches that one can make narrow beams for permitting intensive frequency reuse and the objectives of minimizing path loss.

These techniques can generally be applied in parallel and therefore are not certainly mutually exclusive. Thus, the "best design" for the forthcoming space antennas might include a combination of these different strategies (Winn, 2002; Angeletti *et al*., 2008).

1. Use of higher frequency antennas along with a smaller RF wavelength. If one meets higher frequencies and hence smaller wavelengths the productive ability of a satellite antenna and its "gain" turn around. The antenna's aperture can be shorter with a higher frequency. Because the spacecraft antennas are receiving and transmitting shorter wavelength signals, a smaller antenna can obtain the same outcome as another larger spacecraft antenna performing at a lesser frequency and therefore a larger RF wavelength. Undoubtedly, because the "gain" of an antenna and the square of the wavelength are inversely proportional, this makes a striking influence on the demanded size of the antenna required to attain the same persuasive performance. Today, the largest satellite antennas are available for mobile satellite communications, and the aperture size of such kind of antenna is overturned to a larger dimension as the range of the antennas for down-linking signals to mobile users are from 1,700 MHz to 2,500 MHz. These "lower radio frequencies" are partially utilized because there is no need to have a direct line of sight (LOS) of the signals to the satellite and can achieve the link without inevitably having to "see" the user station that could be blocked partially by a telephone pole or the top of a car (Wakana, 2003).

   The negative aspect of this examination is that the antennas working in these mobile satellite frequencies in the UHF and L band frequencies require to be greater to configure for the transmission of the larger wavelengths. In comparison, the satellites that consume higher frequencies, e.g., Ka-band (30 and 20 GHz) need a direct LOS, but at these frequencies, the spacecraft antenna required to obtain similar gain permits an aperture size, particularly on the order of 100 times smaller. This is true from when the frequencies are 10 times greater and hence the wavelengths are 10 times shorter. As the aperture size is compelled by the square of the wavelength, the aperture size computes to be 10^2 or 100 times tinier. However, in the lower bands, the deficiency of available bandwidth is forwarding satellite systems working toward these greater frequencies for services besides mobile satellite communications, i.e., broadcast, or fixed satellite services (Gritsuk et al., 2016; Hu et al., 2019).

Sadly, at these higher frequencies, there are dilemmas with rain and other kinds of precipitation attenuation. This needs higher link margins to transform into demands for either larger and higher gain antennas or higher power. Additionally, electronics technology is much more challenging in terms of needing the generation of quite high frequencies and tiny, quite precise wavelengths. This leads to higher costs (Plaza et al., 2010; Makarova et al., 2018).

Even though the antenna's aperture can be shorter, its contours should be very much exactly shaped to operate the smaller wavelength beam in a quite accurate way. In addition, this exacting contour also leads to higher costs of antenna fabrication. Additionally, the satellite should be very much exactly pointed to the Earth also therefore, the beams can be targeted more precisely. Briefly, while the shorter wavelengths and higher frequencies permit the spacecraft antennas to be smaller, the complexities just mentioned can more than negate the benefits of the shorter aperture size and turn into greater manufacturing costs. Techniques to focus on precipitation attenuation, exacting manufacturing techniques, and new electronics technologies for the EHF bands are all incorporated to make rise the costs. However, ultimately, these challenges of migration to use these higher frequencies and new bands are reduced. The costs turn to decrease from when the ground systems and progressively satellites are manufactured and installed in these new bands. The next horizon will be the next frontier since the Ka-band systems are used, which are the frequency bands in the 38 and 48 GHz bands (Godara, 1997; Buticchi et al., 2017).

2. Phased array antenna technology. There is a novel technology that is appropriate for the formation of larger-scale spacecraft antennas to enhance ultimate satellite performance. This technology is known as phased array antenna systems. An array of electronic components is linked with this kind of antenna to create highly focused antenna beams and "virtual" high performance. This kind of antenna can facilitate highly efficient multi-beam transmissions. The consequence is an antenna system that permits the effective reuse of available RF frequencies as many times as one can. This technology can be used in two ways. One way is merely by directly spanning today's satellite technology. This

way persists to deploy a very huge high-gain traditional antenna reflector but utilizes a phased array multi-beam feed system to produce a very large number of beams via reflecting-off of an immense parabolic reflector (Iida and Pelton, 2003).

3. As described above, this could be an "untethered" or "tethered" large-scale reflector. The feed system could utilize a horn feed array or a phased array feed using more traditional technology.

4. The other and more technically advanced approach to proceed would be to form a "phased array antenna where different electronic components actually 'electronically form' a beam for reception from a 'virtual reflector' or transmission." A virtually shaped electronic beam of large-scale dimension and of arbitrary shape can be created with this more advanced technology. As in this situation, the beam is "virtually created," the effective size of the beam can increase pretty large by moving from around a 'six by six' phased array to a 'twelve by twelve' or indeed a 'hundred by hundred' phased array. In theory, 10,000 various "pencil-thin" beams could be generated by a 'hundred by hundred' phased array which allows over and above 1,000-fold reuse of the identical spectrum band applicable for satellite communications. The drawback of this approach is that because still the technology is at a very initial stage and formation of a phased array antenna of such kind is quite costly. The expansion of phased array antenna technology could possibly go utterly far. One solution is to set up a large number of phased array elements into space being a free-flying cluster. A cluster of distributed "picosatellite array components" could be formed by the micro-elements of a "virtual antenna reflector" encompassing maybe square kilometers and generate beams that would make "picocells" on Earth (Iida and Pelton, 2003).

The capacity to reuse RF frequencies with such a device might move up to nearly hundreds of thousands of times. Such concepts are simply that at this time. There would be several technical issues to be solved. These would involve the problem of how to regather all of the phased array elements-may be with magnetic attraction so as to reduce orbital space debris (Figure 8.1) (Aumann et al., 2003; Yu et al., 2009).

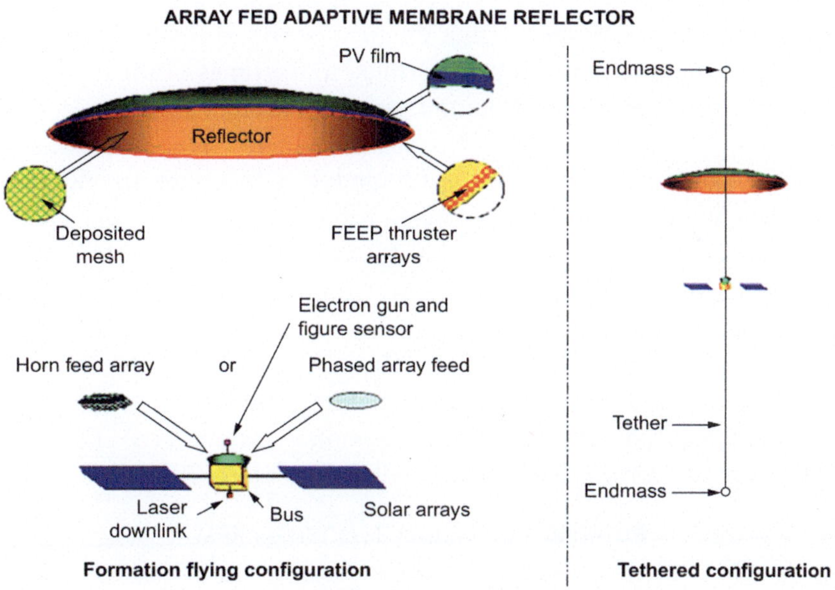

Figure 8.1. *Array fed adaptive membrane reflector; untethered or tethered.*

Source: https://link.springer.com/referenceworkentry/10.1007%2F978-1-4419-7671-0_24.

There are also challenges of interference with other terrestrial communications systems and satellite systems that use identical frequencies (Bright *et al.,* 2018).

There are alternate technologies that might be operated to upgrade forthcoming satellite communications performance also. "Hopping beam" or "Scanning beam" technology is one of the advanced approaches that possibly be used. This is a sort of technology that favors a number of spot beams at a state-of-the-art satellite to operate dynamically in the time domain with code division multiple access (CDMA) or time division multiple access (TDMA) multiplexing. In this sort of antenna configuration, beams can be aimed at various locations, and bursts of data of differing spans (evaluated in milliseconds) can be transmitted to different stations depending on the situation of traffic demand (Phadke, 1993; Lindström and Thornell, 2009). In this technology, streams of broadband traffic (e.g., data, video, and voice) is allowed to be sent in a burst to a specific location shielded by a spot beam and after that "hop" to the afterward location, and next "hop" to another destination, and so on at very short time intervals. The benefit of this sort of "hopping beam" is that the time period of the broadband burst of the digital data stream can be controlled to the times of the day because peak loads differ from time zone to time zone. Moreover, if the satellite is working in the future "W" or "Q" bands (38 and 48 GHz), or in the higher frequency bands, e.g., the Ka-band (20 and 30 GHz), then the dwell time of the data burst in a specific spot beam can be controlled to take care of a heavy rain story or other types of rain attenuation. For data transmission in a specific beam, a dwell time of possibly 10-ms perhaps is doubled or even tripled in a span in a location where the rain rate is heavy and therefore be a reason for severe rain attenuation. These kinds of ideas were tested in the US experimental communications satellite program ventured by NASA called the advanced communications technology satellite (ACTS) (Figure 8.2) (Beukes and Enslin, 1993; Dheepadharshani et al., 2019).

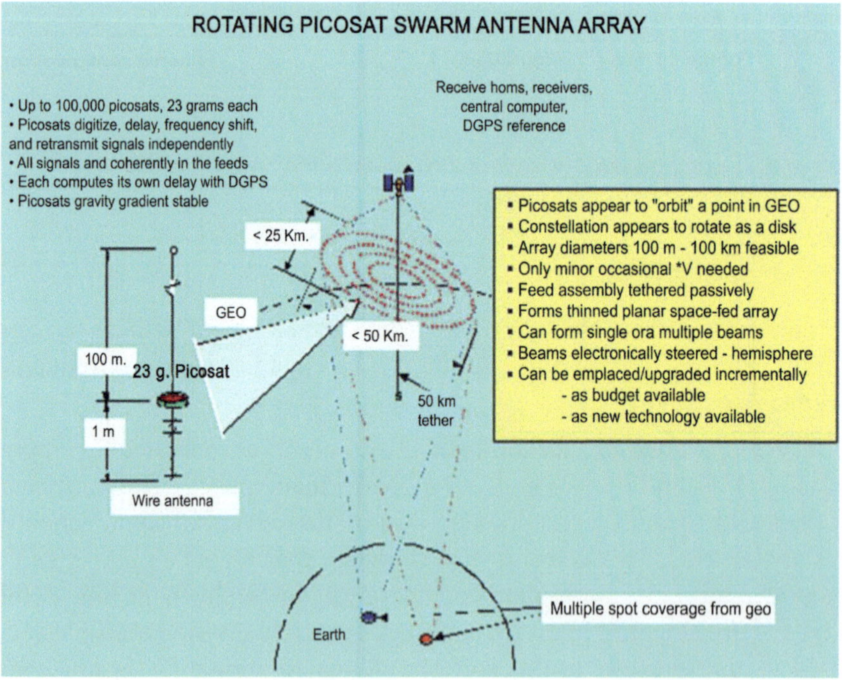

Figure 8.2. *Abstract scheme of a Pico satellite array free-flying in space.*

Source: https://link.springer.com/referenceworkentry/10.1007%2F978-1-4419-7671-0_24.

Despite the specific forthcoming antenna designs, the problem will likely be to attain the competency to produce more and more directed spot beams that can grant insignificant path loss and better frequency reuse by possessing transmission beams that "spread out" less within the Earth and satellite. The key will be to consolidate intelligence along with the enhanced antenna systems to make the antennas operate more efficiently. This could be boosted "intelligence" that would be used on "beam hopping" competency so that digital transmissions in specific beams could better be coordinated to actual comprehensive demand for transforming peak load needs, communications services, atmospheric situations, or other needs (Fleischauer and Hilton, 1988; Subudhi and Pradhan, 2012).

8.3.2. Improved Transmission Systems and Onboard Processing Systems

The key to the effective output of communication services depends first and foremost on improved digital processing potential. The satellite communication systems' efficiency can be evaluated straightforwardly in terms of digital throughput or simple in bits/Hertz. Techniques, for instance, polarization discrimination, the interconnection of geographically separate spot beams, and working at higher frequency bands while wider spectrum bands are assigned to satellite communications support a satellite to grow its amount of accessible bandwidth. Digital encoding (and particularly more efficient decoders and coders (Codecs)) and enhanced digital processing, multiplexing, and modulation techniques favor more bits to be transmitted per Hertz. A few years back, a traditional communication could offer 1 bit per Hz and could utilize a range of strategies to reuse available spectrum by nearly 6 to 8 times. Now, by the use of advanced codecs communications satellites can deduce nearly 2.5 bits per Hz and can reuse spectrum along with factors on the order of 20 times or higher-especially in the context of mobile satellite systems that work at the lowest frequencies and hence have the lesser amount of available spectrum (Reynolds et al., 2013; Levchenko et al., 2018).

However, the trends for terrestrial fiber optic connections and satellite communications will pursue various patterns for particular technical reasons that separate how terrestrial cable and communications satellites work. There are two key benefits in the case of fiber optic networks. The fiber optic networks work, rather than in the RF frequencies, but in the really higher light wave zone of the electromagnetic spectrum. There is an amazingly large volume of spectra available for transmissions in the optical wavelengths and (Dense Wave Division Multiplexing) presently functions in these bands through multiplexing signals just a quarter of a nanometer apart to obtain immense broadband throughput speeds. Moreover, there are very few sources of interference or external noise in the fiber optic cables, and therefore there is a high quality of the signal over somewhat protracted transmission distances.

This means that because of low noise and broadly available spectra, fiber optic networks do not have to be rough as concerned with communication efficiency as is the situation with satellite networks (Godara, 1997; Aumann et al., 2003). Additionally, since satellites should interconnect downlinks and uplinks and interconnect increasingly various downlink and uplink beams, time-based multiplexing systems needed to be used by the satellite systems; therefore there is time for digital processing linked with these switching operations. Fiber optic networks almost operate particularly with "wave division." This is partly because for processing purposes, fiber transmissions do not need "time division" intervals. In contrast with satellite networks that must confront the issues of time delay spoofing, beam interconnection, etc. These conflicts in multiplexing techniques are significant because these competing ways contribute to separate the satellite world and the fiber world. However, the terrestrial wireless world of telecommunications has analogous constraints, especially in terms of multiple reuses of the same frequencies frequently and beam interconnection (Bright et al., 2018). Hence, the similarity of terrestrial wireless networks and satellite (incorporating their reliance on processing time for "cell" or "beam" interconnection), and the ever-growing global demand for mobile services aids to tie the satellite networks and the terrestrial wireless together. This analogous way to time-based multiplexing, as utilized by terrestrial mobile and satellites, operates to make sure that the forthcoming standards for the interconnection of terrestrial wireless, fiber optic networks, and satellites will remain consistent protocols for universal global communications linkages. Undoubtedly, a forthcoming challenge will be to maintain all kinds of telecommunications transmission media to pertain as "compatibly and seamlessly" as possible (Sachdev, 2004; Ellery *et al.*, 2008).

8.3.3. Improved Satellite Power Systems

There were numerous elements of the story of improved satellite power systems. Foremost solar arrays have tremendously grown in size. For these arrays, deployment systems have developed more delicately to permit these many large-scale systems to unravel or, in the other case, be deployed from the short configurations needed to adjust in the rocket fairings at commencing. Next, the "efficiency" of PV solar cells performance regarding conversion of solar energy into the power needed to produce RF signals has also been raised. Solar cells have upgraded from vague silicon solar cells to structured silicon towards gallium arsenide cells. In addition to that, the number of junctures or gates where solar energy is caught has raised and upgraded to the ultraviolet side of the spectrum, where the greatest amount of energy is achievable. Briefly, the efficiency of energy conversion has raised from nearly 7% to 30% and promptly may become almost 50% in the remarkably efficient systems. The performance of solar cells has improved and also the potential of the satellites to show PV cells has also improved so as to obtain maximum solar radiation. The revolution from having solar cells firmed on the exterior of cylinder-shaped satellites at the location where the sun was "hidden" 40% of the time to 3 axis body balanced arrays has created a great difference. Currently, the solar array can be uniformly directed toward the sun and indeed "angled" to obtain the maximum illumination. Rather than when the satellite along with its solar arrays is in Earth eclipse, the arrays are now installed with the great efficiency to ingest the most solar power that is feasible (Brown and Eremenko, 2008; Yin *et al.*, 2016).

The increase in battery performance has the third leading trend. There has been a boost in the "energy density" of the battery systems and in their viable lifetimes. From Nickel-Cadmium batteries to Polymer Lithium-Ion cells, the batteries have raised in performance. The lifetime of solar cell arrays and batteries-both of which fade in performance in orbit-are critical elements in the capacity of the spacecraft to carry on to operate over protracted periods of time in the nasty environment of outer space. From nearly 100 W have been increased by the power of communication satellites together with the Early Bird (or Intelsat I satellite) to power systems that produce on the order of 15 kW. Shortly, thorough output power performance has raised by a factor of approximately 150 times. There are also many protracted lives of these power systems. However, these systems are also very much massive. If the net-performance is to be evaluated in orbit for satellite power subsystems in watt per kilogram per year, the net raise is high on the order of 10–20 times (Figure 8.3) (Chai Ji *et al.,* 2019).

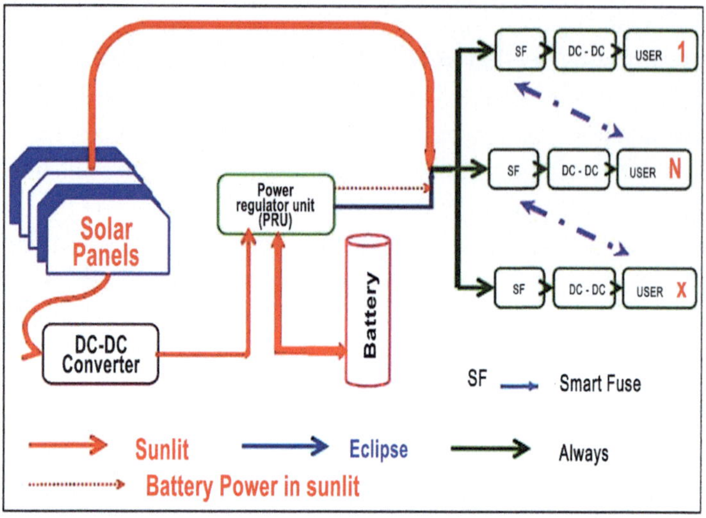

Figure 8.3. *Diagrammatic scheme of satellite power system.*

Source: https://directory.eoportal.org/web/eoportal/satellite-missions/p/pisat.

The promising improvements have been seen in the future. A net efficiency of over 60% might be able to obtain by the so-called rainbow solar cells that have possibly seven various PV junctions. These standards of efficiencies might also be able to obtain by the so-called quantum dot energy systems. Even yet both of these technologies are in the laboratory, and therefore not still accessible in the commercial market (Ippolito and Pelton, 2004; Liang *et al.*, 2007).

The quantity of solar power obtainable in space is tremendous in case that it could be captured effectively. One idea is to have some kind of concentrator or solar collector to brighten solar cells through higher intensity. There can be quiet mirror surfaces connected to a solar array; thus, the cells can look at the proportionate of two or three suns. Nevertheless, there are more advanced ideas that would install very lightweight film-covered collectors that could brighten solar cells by great intensity. This technology is being created in combination with designs to develop solar power satellites (SPS) so that the equivalent of hundreds or even thousands of suns could be seen by the solar cells. However, this same technology could be deployed in combination with a large-scale communication satellite platform. These large-scale and much low mass solar concentrators could be planned and deployed at really less cost as compared to high-performance solar cells (Figure 8.4).

There have been different researches to test whether battery- and solar-powered satellites have the potency to "peak" in performance, concerning maximum lifetime and power. These studies look upon to examine in contrast battery systems and large-scale solar arrays vs other power sources like regenerative fuel cells, nuclear energy, etc. Various projected results and optimization formulas have been generated by these assorted studies, although many thinks that for power systems beyond 20–40 kW, regenerative fuel cells or nuclear energy might turn out to be more cost-effective (Sheard *et al.*, 2012).

Undoubtedly, isotope-based SNAP generators have been applied by various space projects for high-powered and long-term missions, but safety is always concerned (both at reentry and launch) with nuclear power sources. For the development of nuclear reactors using ionized gases or thermal, there are more determined longer-term research projects to support propulsion systems. In these situations, to produce electrical power, nuclear power could also be used, which facilitates a lot of missions in the space applications field. Still, such systems are in the process of development by various major space agencies. There are development challenges associated not only with usable lifetime, and mass-to-power performance ratios but prominently with nuclear generators in space increase to deal with the security of radioactive materials and their secure disposal (Massari and Zamaro, 2014).

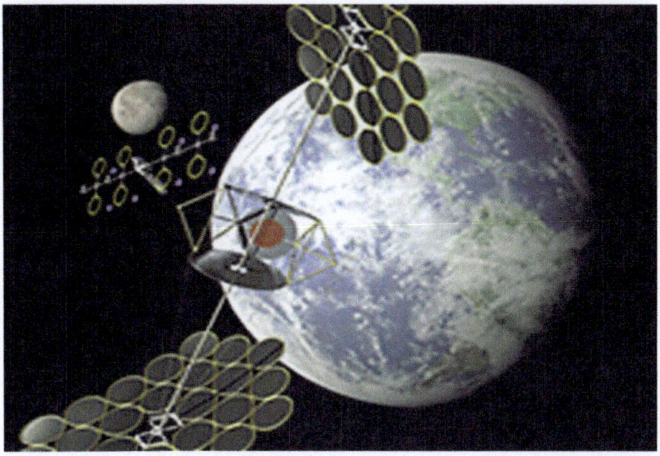

Figure 8.4. *Design ideas from Nasa for lightweight solar concentrators.*

Source: Image courtesy: NASA; https://www.livescience.com/solar-power-stations-in-space.html.

However, there are some technologies having great potential that do not include the dangers of nuclear power. Two of the most auspicious are regenerative fuel cells that can generate reliable power for protracted periods of time using high mass-to-power ratios and adequate to function independent of sun exposure. Current development to utilize fuel cells for Earth-bound energy needs for cars and buildings advises that these may be used effectively in space within the 10 years ahead. The systems are even involved in more immediate space-based energy systems that create additional efficiencies using solar power. One novelty is that of many lightweight solar concentrators that be utilized not only using SPS but as a way to focus the equivalent of radiation of numerous suns on state-of-the-art solar cells or alleged quantum dots. The ability to be possibly three times more efficient in transforming solar radiation into electrical power is promised by the quantum dot technology that is still in the laboratory. A quantum dot can be stated as a unique type of semiconductor whose "excitons" are restrained from roving in any dimension. Quantum dot units' characteristics are given by this constraint eventually that lies within those of traditional semiconductors and the role of an individual molecule. A broad range of applications is being sought by nanotechnology for "quantum dots" in energy systems, medical imaging, and other fields. The use of near-infrared and may be higher frequency quantum dots being a retrofit to currently silicon solar cells might be involved by the nearer term applications regarding quantum dot technology to improve performance.

8.3.4. More Effective and Reliable Spacecraft Design

The principal drivers of satellite communications design and performance will probably be digital encoding and increased power, antenna design, and multiplexing techniques that will permit higher throughput. However, a good design for the spacecraft is also significant to keep up design features and reliable long-term operations that can make it able to manufacture the satellite more speedily, at lower mass concerning lower-cost launch or other kinds of improvements (You et al., 2015).

8.4. SATELLITE ORBITAL CONFIGURATIONS AND IMPROVED SPACECRAFT ORIENTATION AND POINTING

The following trends can chiefly summarize the previous few decades in the development of satellite communications. These have been to create spacecraft systems stronger and to install larger aperture antenna systems with much higher gain and the capability to reuse available spectrum using the interconnection of cellular-like spot beams. The gains have been dominated in satellite capacity by these advancements plus the proficiency to launch and install larger spacecraft in space plus the gains that digital compression and digital communications techniques have carried to the satellite communications industry and permitted various new applications to emerge, especially in mobile, fixed, and broadcasting services. Even various kinds of smaller satellites exist differently described as nanosatellites, micro-satellites, pico-satellites, etc. Regardless of the innovative use of digital processing methods, these spacecraft depict far less than 1% of in-orbit capacity (Smyth *et al.,* 2010). The ability to give consumers low-cost transceivers and the forthcoming extension of satellite system capacity seem to need the deployment of multi-beam, even larger aperture antennas in space. The issue is that this may be somewhat arduous to obtain unless one starts to progress toward one of the several choices with contemporary and projected launch capacity. These would be: (i) the installation of "parts" of systems that are either composed in low earth orbit (LEO) and after that drifted into Geo orbit to forge large-scale satellite platforms using antenna apertures in an overabundance of 30 m; (ii) the formation of "networked" antenna systems that take off in some sort of formation or are connected together to form a "virtual antenna system"; or (iii) the production of a large-scale constellation managing the extremely high frequency (EHF) bands using a really large number of satellites in the constellation (i.e., identical to the proposed Teledesic satellite network). The previous section of spacecraft antenna presented some of these concepts (Baker *et al.,* 2011). In spacecraft design, there are always different kinds of trade-offs in terms of lower cost, optimizing system capacity, and smaller user antennas, techniques to manage problems like precipitation attenuation, and system lifetime. One of the principal constraints that would lie using very large antennas having apertures over 30 m is that there would be a need for precise pointing accuracy of the space antenna. There would be a need to take into consideration such aspects being a thermal expansion of the antenna because of the exposure to solar radiation, etc. In these scheme trade-off considerations, it is obvious that the space antennas are allowed to be smaller in size by enough higher frequencies in the EHF (i.e., 30 GHz and exceeding), but on the other side, the challenges of precipitation attenuation (particularly rain) appear to be much more drastic and enormously obscures keeping user antennas on the ground low in cost and small in size (Falkenmark and Rockström, 2006).

8.5. NEW GROUND USER SYSTEMS

The prevailing trend in all forms of information processing, digital communications, and digital entertainment (i.e., the communications, information, and entertainment enterprises, or the ICE industries) have been to prosper consumer-oriented distributed systems that have moved nearer and nearer to the edge. This means that entertainment systems, computers, and communications devices have become more compact, smaller, lower in cost, and more user-friendly so that typically consumers can keep and use these devices. This same trend has been followed obviously by satellite communications. In the 1960s, computers were highly expensive devices, massive, and highly centralized that demand a team of specialists to operate. The same was for the satellite earth stations. Besides the cost of $10 million (US) of a classic Intelsat Standard an Earth Station of that age, it also demanded a team of 50 or more experts to operate and participated in the accurate pointing of a 30 m (93 ft) antenna that weighed a lot of tons. In contrast to the first digital computers, today's handheld computers have much higher computational power. Similarly, handheld satellite transceivers are sold straight to the consumer that can exchange information to in-orbit satellites straightforwardly. The cost of the consumer video games, digital satellite phones, and computers is in the hundreds of dollars, and it also demands minimum training for its consumers to employ these products (Esch *et al.,* 2017).

This trend toward digital instruments that can be obtained at low cost and miniaturization of user-friendly consumer devices will uncertainly continue. This means that the trend will proceed toward even more compact "wearable devices." Considering that health-related problems, including RF radiation can be overcome successfully, it seems viable that the later tier of development might be represented by the embedded communication devices that have a capacity of linking to in-orbit satellites (Frey and Smith, 2007; Gong, 2007).

There are obviously a lot of technical challenges to face and reduce. The population continues to grow rapidly with some 6.5 billion people on the Earth and, each day, more and more bodies are looking for access to broader band communications to assist video messaging, entertainment, and data and voice communication. Moreover, a large number of people are looking for broadband mobile connections either through satellite communications connections or terrestrial mobile services. New solutions are required by this increasing demand for broadband mobile services. These can be in the form of new techniques to reuse available frequencies too intensively within satellite or terrestrial wireless systems, more efficient digital compression strategies, the movement to higher frequency spectrum bands, or some combination of these solutions. All these endeavors become more challenging by the smaller consumer devices having smaller antenna size. Application specific integrated circuit (ASIC) devices have participated largely to the capability to underestimate consumer handheld communications devices, but a discovery in quantum computing will be required to obtain new levels of miniaturization. For the reduction of power requirements, these quantum computing level breakthroughs may aid as well. This devaluation of power would assist not only with the issue of portable power supplies but also would help with health-related matters (Figure 8.5) (Evans et al., 2005; Hagar, 2007).

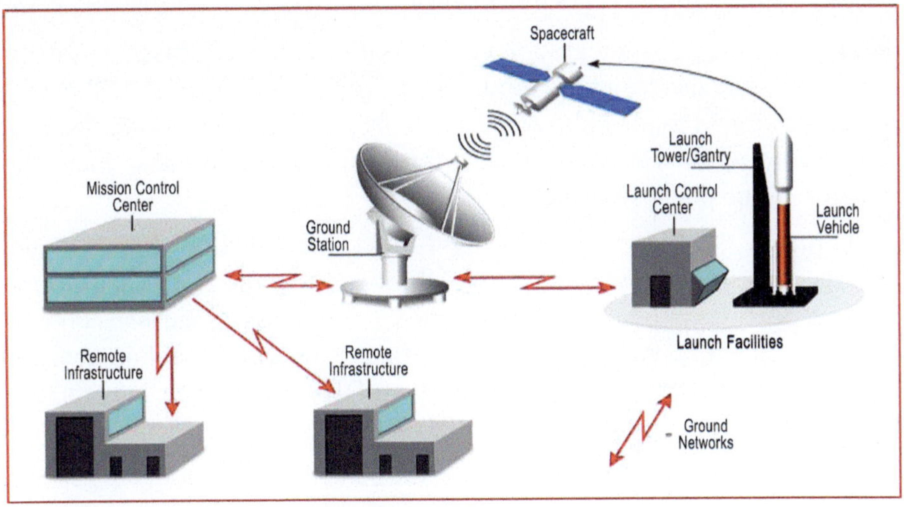

Figure 8.5. *Satellite ground systems.*

Source: https://ai-solutions.com/newsroom/what-are-space-ground-systems/.

8.6. INTEGRATED SATELLITE AND TERRESTRIAL SERVICES AND NEW MARKET DEMAND

Breakthroughs in the next generation of ASIC technology, and quantum computing nano-material engineering can all assist to promote the development of forthcoming communication devices utilized in satellite-based networks and also terrestrial mobile systems, but the key may also be obtained in the integration of mobile systems and terrestrial cable with satellite networks in a novel and innovative approaches. Briefly, the leading keyway, the continuous expansion of broadband services, and the given frequency constraints would seem to associate using fiber optic systems for intra-building risers, terrestrial wireless for localized services, "trunking interconnections," and urban-wide area networks, and to use satellites for regional and global mobile services, for broadcast applications, and for services in remote, rural, and island areas (Navis and Glynn, 2010).

Seamless digital networking standards are the principal to this sort of integration of broadband services-presumably based on internet protocol (IP) standards. Satellite communication networks, fiber optic cable systems, and terrestrial wireless all have their potentials, and seamless interconnectivity favors all of these transmission media to be optimized. An intellectual platform will be allowed by the merger of IP-based protocols on a worldwide basis upon which this integration of different sorts of transmission systems will be possible increasingly. Similarly, the TCP/IP and Internet will permit for all the space-based application satellites to be combined as applications that can be approached using "smartphones," of course, also including "smart satellite phones" (Chien *et al.,* 2005).

Formerly satellite communication links, particularly those to and from geosynchronous satellites, located around a tenth of the path to the Moon at 35,870 km (or 22,230 miles) raised to the Earth's surface confront major issues with communications operating using TCP/IP due to delay or transmission latency. The actual design of TCP/IP aimed to interconnect PCs on the Internet (Ogilvie et al., 2008).

Detected delays were assumed to be the outcome of network congestion, and the links were time out automatically to switch to recovery mode. Several alterations have been made over time to optimize satellite working by TCP/IP. There particular IP over Satellite (IPoS) standards that apply reset of timers to fit satellite delay, "spoofing," and other methods to allow satellite links to work at more high efficiencies. Additionally, for the accommodation of virtual private network (VPN) security measures and IP Security (IPSec) when utilizing satellite links, changes have been made. Piece of the problem is that various techniques have been developed by the International Telecommunication Union (ITU) and the internet engineering task force (IETF) to obtain efficient network interconnections. For the satellite industry, part of the forthcoming challenge is to be capable to accommodate IETF and ITU requirements and standards effectively, economically, and rapidly

(Koren *et al.,* 2018; Norgren *et al*., 2015).

8.7. TRACKING, TELEMETRY, MONITORING, COMMAND, AND AUTONOMOUS OPERATIONS

In the future, it seems presumable that satellite systems will go on to become more sophisticated and larger. As this progress continues, more complex roles will be tended to assume by the communication satellites concerning onboard signal processing, onboard switching, and other operations that were once operated on the ground exclusively. This metamorphic movement, specifically toward onboard processing, will form satellites more adept and better able to manage rain attenuation, inter-beam connection, and other advanced operations (Di Silvestre et al., 2020). This trend toward higher complexity in space will create the role of tracking, telemetry, monitoring, and command more arduous. Particularly, the engineering and software required to obtain speedy fault detection when there possibly be something like a 1000 beams and millions of likely inter-beam connections turn to become massive (Adushkin et al., 2016; Skinner, 2017). The most challenging role for the coming years in terms of engineering progressive communication satellites will be the making of computer code to speedily discover a specific defect in onboard link interconnections. For these forthcoming TTC&M roles, particularly detection of interference and fault detection, there will be a trend to apply artificial intelligence to aid with these operations. Similarly, there will be growing endeavors to apply artificial intelligence and computer programs to manage as many of the functions of the satellite up to its 10–18 years' lifetime so-called "autonomous operation."

Communication satellites work 24/7 over the entire year and are managed by a team of technicians and human engineers to keep an eye on all of these processes and to interconnect in rapid interference or fault detection which is uneconomic increasingly. Thus, for satellite system design, one of the tremendous technical threats of the future will not only be the onboard processing and the progress of complex multi-beam antenna systems but every new kind of onboard intelligence that this will indicate. Hence, it seems that the forthcoming communication satellites will have telemetry, largely automated tracking, monitoring, and command, and AI-based autonomous operations and fault detection systems. This signifies that human-originated commands will be activation of redundant receivers, the exception and battery discharges, recording of billing information, a shutdown of nonfunctioning switches, and hundreds of other functions that were once regulated by ground operators continuingly will evolve into tasks that are assumed increasingly by onboard computers (OBCs). However, there will yet be a requirement to uphold satellite operations centers and monitor many satellite operations. However, artificial intelligence and autonomous operations will facilitate to avoid these centers from increasing size exponentially and hence aiding to create uneconomic satellite operations (Wasser, 2011; Malinowski *et al.*, 2017).

8.8. FUTURE TRENDS FOR MARKETS AND REGULATORY SYSTEMS

The future progress of satellites does not rely on only technological development. Regulatory actions and market demand are similarly substantial drivers of the satellite communication industry. Present's market trends recommend that novel growth will be devised by demand mainly in the three prime areas. These are: (i) Mobile communication services in areas that are not well marked by terrestrial wireless services (i.e., maritime, air, and remote land areas); (ii) broadcasting and entertainment services (i.e., high-definition, and 3D digital television, video, also broadcast radio that is connected with emergency vehicular services also); and (iii) Holes in communication services that are not well marked by terrestrial wireless broadband systems and fiber optic networks. In various developing regions, still satellite broadband remains principal for Internet connections. New kinds of Internet-optimized satellite systems like O3b (the Other Three Billion) are tailored to bring Voice over IP and broadband Internet services to the areas of the world where still there is a lack of effective terrestrial data and telephone networks (Fichman et al., 2011; Oteman et al., 2014).

This is not to propose that there might not be alternate market niches for satellite communication systems. One such additional satellite service is represented by store-and-forward satellite systems that facilitate business-to-business (B2B) services and messaging services and are sometimes also bound to space navigational services for trains, trucks, ships, and buses. One of the mysteries about forthcoming market requirements corresponds to what should be called integrated space applications. In the era of "smartphone" applications, it seems progressively fair that applications to assist promptly weather data updates, interactive navigation, and remote sensing applications will emerge over time. Today Earth observation, remote system, space, and meteorological navigation systems are brought via separate space-based satellite systems and the transmission of information is mostly via "stovepiped" and separate telecommunication networks (Gubic and Baloi, 2019). In the future, these systems can and possibly will be joined through Internet linkages to become just supplemental applications available using handheld "smart devices" or eventually possibly through embedded chip technology. All of these alterations will assist to transform the structure of the satellite communications industry. Traditional industry divisions will be broken down by acquisitions, mergers, and market integration through Internet applications. This will signify at one level that organizations will merge across transmission technologies, e.g., cable television, fiber, satellites, and terrestrial wireless. On an alternate level, organizations in one space-related service like satellite communications can and possibly will be varied into other space applications, e.g., space-based messaging, space navigation, real-time situational awareness, and remote sensing (Lyon, 1998; Pecorella *et al.*, 2015).

Including service and market demand, a critical part will also be played by the regulations. One of the most conspicuous areas will be that relevant to the regulatory addition of new abilities (i.e., novel RF allocations that could hamper with satellites, i.e., those yielded for high altitude platform systems (HAPS)), and frequency allocations. Such HAPS may be installed over urban regions to give remote sensing, television, or wireless broadband communications. Today for new satellite communications networks, the deficiency of new orbital locations in GEO orbit, and the deficiency of available spectrum are involved by one of the leading constraints to the extension of satellite communications services (Dunbabin et al., 2006; Kalita *et al.*, 2018).

This problem will be provoked by the demand for increasingly terrestrial wireless services and expanded broadband services to support mobile applications. This will derive technology to create enhanced ASIC transceivers to work more effectively in this spectrum-limited environment and the larger multi-beam satellite systems to make rise frequency reuse. Today a key role is played by the ITU in the adaptation of spectrum and the establishing of suggestions to restrict intersystem interference. The ITU is restricted in a role in various ways. Nations often enact via footnotes restrictions on frequency spectrum allocations not over their national borders. There are no exceptional enforcement powers of the ITU, e.g., penalties or fines for those who do not completely deploy its proposal. In the future decades, the growth of the satellite industry could be limited in a serious manner by spectrum shortages and given the deficiency of saturation and enforcement powers of orbital locations (Campbell *et al.*, 2009, 2013). The growing spread of orbital debris is another severe regulatory issue. This spread is especially agitating in the LEO, but more it is a concern as well in the GEO orbit and the medium earth orbit (MEO) also. Controls designed by the UN Committee on the Peaceful Application of ITU and Outer Space and national endeavor at expected diligence in these regions are initiating to have little favorable influence, but the persisted deployment of new systems may also highlight this problem as a potential impedes on the future advancement of the satellite communications industry and alternate space applications. Using advanced high-definition tracking systems beyond 30,000 objects, the human fist-size can be traced in Earth Orbit. The anticipation is that high due diligence endeavor to reduce sources of orbital debris and deteriorating materials that fall from LEO can lead to the issue of "space junk" under control prior to cascading effects occurring from orbital collisions can make a blizzard of dangerous materials in space (Saunders *et al*., 2004; Hu *et al*., 2019).

Currently, there are efforts to make a worldwide database to follow the orbits of satellites initiating with the Geo ring where many communication satellites are already in operation. SES Global, to date Intelsat, and Inmarsat have admitted to input data, and Echostar, Eutelsat, and Telesat have pointed intent to contribute (Chan and DalBello, 2010; Altaei and Mhaimeed, 2017).

8.9. NEW INITIATIVES IN SPACE COMMUNICATIONS

Satellite communication technology and operations still indicate a too speedily emerging field. A variety of technologies are driven by the telecommunication satellite industry to make better user equipment and space systems but also by challenging communication systems-especially fiber optic networking. The curve is also being driven by other technologies. Hence, the future of satellite communications will also be transformed by robotics, artificial intelligence, HAPS, new multiplexing systems, terrestrial wireless systems, laser communications, quantum computing, and a host of other technologies. Eventually, the future of mankind is based in space, and therefore space communications will continue to be a part of the future one way or another. If the development of satellite communications is considered in terms of power, throughput capabilities, lifetime, antenna gain, and costs, it has upgraded by a factor of over 1000 times in the back half-century, and there is an abundant technology in the pipeline that it could upgrade another

1000 times in additional 50 years. However, the key to seeing the future is not by protruding the rate of technical innovation but to find to understand fundamental market trends and to understand what types of applications the public will demand to meet future environmental, societal, and economic requirements in the future decades. Entirely new markets could emerge well (Ejaz *et al*., 2018; Yang *et al*., 2018).

8.10. KEYS TO THE FUTURE OF SATELLITE COMMUNICATIONS

Anticipating the future is always complicated, but there are obvious indicators as to opportunities and trends. Currently, the following "keys" are noticeable.

8.10.1. Off-World Communications

The most solely adapted new market for satellite communications appears presumably to be cislunar communications to Moon-based colonies also connects to inhabited asteroids or Mars, the satellites revolving other planetary bodies in the solar system, or artificial space colonies. The technology to endure communications with the assistance of current scientific satellites probing already the Solar System will give an advantage of an early start in this direction. Laser-based communications would be seeming like the most logical expansion of potential in this respect, but the deficiency of an atmosphere on the asteroids, Moon, or the satellites of diverse planets causes light communications reasonably viable at the large distances of the Solar System. Laser beams are well aimed and hence are very less sensitive to path loss because of spreading. The ability to minimize path loss is imperative if the transmission distances exist millions of miles (Lopes *et al.,* 2011; Burleigh *et al*., 2019).

It would be suggested by the light attenuation inside the Earth's atmosphere that laser communications would be conducted toward Earth-orbiting satellites, possibly in geosynchronous orbit. There have been alternate approaches as to how to establish these links most efficiently. Solar sail-oriented satellite is one of the approaches that could be located or "levitated" higher one or both of the poles; thus, a signal could be transmitted straight to anywhere in the Southern (or Northern latitudes). The numerous space agencies, particularly NASA, have devoted a good deal of research and also modern space communications hardware in interplanetary relays and interplanetary (Ruan et al., 2017; Abderrahim *et al.,* 2020).

8.10.2. Smart Satellites and Advanced Encoding

The rapid development in computer technology that was envisioned by the so-called Moore's Law expected a doubling in scope every 18 months and, generally, in computer performance, this exponential growth has sustained for some 30 years. Since the transformation from analog to digital satellite communications, the satellite communication industry has pursued an identical curve of accelerated performance. Essentially, in fact, modern's communication satellites are "software-defined hardware." Even though these are elaborate devices devised to work in the rough environment of space, today, the communications function is substantially the result of swiftly processing digitally encoded information (Liolis et al., 2019; Alsharoa and Alouini, 2020).

Digital processing is allowed to be increasingly efficient by the advanced coding capabilities, e.g., "turbo coding." A few years ago, approximately 1 bit of information per 1 Hz of the available spectrum could be processed by the most productive communications satellites. Today, efficiencies of 2.4 bits per Hz of the available spectrum are obtainable with more efficient modulator/ demodulators (i.e., modems) and more efficient coder/decoders (i.e., codecs). In forthcoming years, the efficiencies of digital satellite communications together with enhanced codecs and onboard processing may be able to obtain efficiencies of 4 or 5 bits per Hz or even improved throughput capabilities (Miao *et al*., 2016; Heng *et al.,* 2020).

The limit on a performance that deduces from the very efficient encoding of information is firmly determined by the amount of interference, noise, or more correctly in digital terms, the bit error rate, that avoids the use of progressively efficient codes. Especially, in the case of heavy rain and related rain attenuation of signal purity, 4-bit encoding might be used, e.g., QAM, but using today's communications satellites, it is impossible to use even greater efficiency 8 bit or 16-bit encoding unless the sky conditions are clear and other kinds of interference does not exist. In the coming years, it is possibly "onboard processing" which can be able to restore uplinked signals to intact quality, and this competency of processing signals onboard the satellite and then as they are acquired as down-linked transmissions can make possible the transmission efficiency to increase (Peng *et al*., 2018; Jia *et al.,* 2020).

It is substantial to note that broadband wireless technologies and satellite systems are most attentively focused on encoding and processing efficiencies, in contrast to fiber optic networking systems. This is for the reason that in the fiber world, there is approximately zero-bit error rate (i.e., virtually zero meaningful interference) and a nearly limitless spectrum using dense wave division multiplexing (DWDM). Under these circumstances, there are no powerful incentives for the increased bit/Hz throughput and the development of high-efficiency encoding. There is limited incentive for progressively efficient usage of available bandwidth using the virtually unlimited available spectrum. In the world of broadband wireless communications and satellites, obviously, vice versa holds true (Hirzinger et al., 2004; Reintsema et al., 2007).

8.10.3. Integrated Satellite and Terrestrial Networks (i.e., The "Pelton Merge")

Today's world of communications is driven to linking various forms of transmission networks. Hence, the target is to integrate "seamlessly" diverse transmission media to provide a variety of consumer requirements. The relief of mobility drives calls for broadband wireless services, incorporating communication satellites and, in upcoming years unattended aerial vehicles (UAVs) and high-altitude platform systems (HAPS) that serve platforms for broadcast services and communications. On the other side, coaxial, and fiber optic cables can provide very high efficiency as well as cost-effective services to defined locations-especially on the involvement of heavy routes of traffic. These divergent communications demand ends in the need for technical and operational standards to favor these wireless networks and wire to connect easily with high quality and at low cost (Fasano et al., 2009; Benedict, 2013).

The goal is easy to understand and define, but still, it is difficult to achieve the "seamless" interconnection. Three key factors stem the difficulty: (a) the world of coaxial and fiber cable multiplex signals within the wave division domain due to the large amount of spectrum that is provided for broadband services, whereas wireless services counting satellites work in the time domain. It is due to the cellular type frequency reuse needed to break down the spectrum into tiny cells in order to make possible the multiple reuses of available spectrum. Digital processing time is required by the interconnection of the signal utilized in these different cells, and therefore, multiplexing is allowed in the time domain. Interconnection is achievable, but the differences add complexity, cost, and technical challenges regarding smooth interconnection. (b) The other impeding factor between the system of satellites and fiber is concerning the need to consume high-efficiency codecs to load more information into the spectrum available in the system of wireless services, as mentioned above. Moreover, "seamless interconnection" is made more complex by the technical differences and hence more costly. (c) The next factor related to the world of the internet that was first designed to execute on terrestrially "wired" networks, e.g., local area networks (LANs). In the initial design, any considerable delay in transmission was considered to be network congestion instead of the transmission delay linked to geosynchronous satellites. Other views of the design, for example, IPSec architecture, were not planned to adjust the transmission architecture of satellites. Originally where IP-based traffic was routed up to satellites, these issues contributed to making very inefficient satellite transmissions, and it becomes difficult to establish secure VPNs due to the IP SEC procedure of "stripping off of header" information to maintain privacy resulted in ambiguity in satellite routing. With the passage of time, more standards corresponding to IPoS were formed, and these problems of compatibility between IP-based traffic and satellite transmission were broadly resolved, by readjusting "clocks" to deal with satellite transmission delay, and other rules to deal with IP SEC needs related to VPNs (Ellery *et al.*, 2008; Woellert *et* al., 2011).

In the coming years, the design of satellite networks and compatible terrestrial to deal with IP-based traffic will be a leading challenge to form wireless and wire networks cost-effective, fully compatible, and of high quality. It is acknowledged by many designers recently that a combination of wireless and wired systems will be required to hold diverse broadcasting and networking needs and particularly the call for mobility. The accomplishment of the so-called Pelton Merge remains still hard to achieve the goal that is based on enhanced interface standards optimized to accommodate the requirements of IP-based traffic, broadband mobile services provided through satellite and wireless systems, and inexpensive fiber optic cable networking design (Sarda *et al.*, 2006, 2010).

8.10.4. Advanced Launch Capabilities, In-Orbit Servicing, and Advanced Platforms

For years, the significant impede on the progress of the satellite communications industry has been regarding the reliability and cost of launch services. Even though satellites have grown in power, capacity, cost efficiency, and lifetime, launch vehicle charges with respect to kilograms of payload to orbit have remained somewhat static. The breakthrough technology is hoped with respect to advanced ion engines, tether lift systems, and electrical propulsion, or in coming years, even space elevators might increase not just the reliability of launch systems, but striking decreases in cost. R&D has also increased on different systems to serve space tugs that might be made possible to propel satellites from LEO to GEO orbit or possibly even satellite service vehicles that could give positioning, new batteries, and orientation fuel or even support new satellite antennas or other elements to in-orbit satellites. Such enhanced space tugs or launch systems could expedite the design and installation of satellite clusters or large-scale satellite platforms that could support largely extended communications capability to orbit also (Burleigh et al., 2019; Adushkin et al., 2020).

8.10.5. Space Safety and Orbital Debris

For ages, the center of attraction of space safety systems and corresponding technology was on human space flight. In late years, nevertheless, there has been a center of attraction on how to make the peaceful usages of outer space better assured and the safer space activities in all its forms. One of the significant steps toward this direction is the UN voluntary guidelines regarding space debris. The current UN COPUOS drive on the "Sustainability of Space" is targeted at focusing on this problem in vast terms that is beyond space debris, and hence the effort is seeking new ways to make sure that the approach to space by all nations can better be guaranteed and finding ways to maintain space from being militarized (Rovetto, 2016; Maclay and Mcknight, 2020).

REFERENCES

1. Abderrahim, W., Amin, O., Alouini, M. S., & Shihada, B., (2020). Latency-aware offloading in integrated satellite-terrestrial networks. *IEEE Open Journal of the Communications Society, 1*, 490–500.
2. Adushkin, V. V., Aksenov, O. Y., Veniaminov, S. S., Kozlov, S. I., & Tyurenkova, V. V., (2020). The small orbital debris population and its impact on space activities and ecological safety. *Acta Astronautica, 176*, 591–597.
3. Adushkin, V., Veniaminov, S., Kozlov, S., & Silnikov, M., (2016). Orbital missions' safety-a survey of kinetic hazards. *Acta Astronautica, 126*, 510–516.
4. Alsharoa, A., & Alouini, M. S., (2020). Improvement of the global connectivity using integrated satellite-airborne-terrestrial networks with resource optimization. *IEEE Transactions on Wireless Communications, 19*(8), 5088–5100.
5. Altaei, M. S. M., & Mhaimeed, A. D., (2017). Satellite image classification using image encoding and artificial neural network. *International Research Journal of Advanced Engineering and Science, 3*(2), 149–154.
6. Angeletti, P., De Gaudenzi, R., & Lisi, M., (2008). From "bent pipes" to "software-defined payloads": Evolution and trends of satellite communications systems. In: *26th International Communications Satellite Systems Conference (ICSSC), 1*, 1–23.
7. Atzori, L., Iera, A., & Morabito, G., (2010). The internet of things: A survey. *Computer Networks, 54*(15), 2787–2805.
8. Aumann, H. H., Chahine, M. T., Gautier, C., Goldberg, M. D., Kalnay, E., McMillin, L. M., & Susskind, J., (2003). AIRS/AMSU/HSB on the Aqua mission: Design, science objectives, data products, and processing systems. *IEEE Transactions on Geoscience and Remote Sensing, 41*(2), 253–264.
9. Baker, J. W., Lin, T., Shahi, S. K., & Jayaram, N., (2011). New ground motion selection procedures and selected motions for the PEER transportation research program. *PEER Report, 3*, 1–33.
10. Benedict, B. L., (2013). Rationale for need of in-orbit servicing capabilities for GEO spacecraft. In: *AIAA SPACE 2013 Conference and Exposition* (Vol. 13, p. 5444).

11. Bergmann, M., Romano, P., Koudelka, O., & Wittig, M., (2009). Generic communication user and system requirements for future space science missions. In: *2009 International Workshop on Satellite and Space Communications* (Vol. 1, pp. 295–299). IEEE.
12. Beukes, H. J., & Enslin, J. H. R., (1993). Analysis of a new compound converter as MPPT, battery regulator and bus regulator for satellite power systems. In: *Proceedings of IEEE Power Electronics Specialist Conference-PESC'93* (Vol. 1, pp. 846–852). IEEE.
13. Bright, J. M., Killinger, S., Lingfors, D., & Engerer, N. A., (2018). Improved satellite-derived PV power nowcasting using real-time power data from reference PV systems. *Solar Energy, 168*, 118–139.
14. Brown, O., & Eremenko, P., (2008). Application of value-centric design to space architectures: The case of fractionated spacecraft. In: *AIAA Space 2008 Conference and Exposition* (Vol. 1, p. 7869).
15. Buehler, R., & Griffin, D., (2003). Planning, personality, and prediction: The role of future focus in optimistic time predictions. *Organizational Behavior and Human Decision Processes, 92*(1, 2), 80–90.
16. Burleigh, S. C., De Cola, T., Morosi, S., Jayousi, S., Cianca, E., & Fuchs, C., (2019). From connectivity to advanced internet services: A comprehensive review of small satellites communications and networks. *Wireless Communications and Mobile Computing, 1*, 1–44.
17. Buticchi, G., Costa, L., & Liserre, M., (2017). Improving system efficiency for the more electric aircraft: A look at dc\/dc converters for the avionic onboard dc microgrid. *IEEE Industrial Electronics Magazine, 11*(3), 26–36.
18. Campbell, M. D., Handley, B., Wise, H. M., King, J. D., & Campbell, M. D., (2009). Developing industrial minerals, nuclear minerals, and commodities of interest via off-world exploration and mining. In: *Paper/Poster at Conference of the American Association of Petroleum Geologists (AAPG), Energy Minerals Division* (Vol. 1, pp. 1–33).
19. Campbell, M. D., King, J. D., Wise, H. M., Handley, B., Conca, J. L., & Campbell, M. D., (2013). *Nuclear Power and Associated Environmental Issues in the Transition of Exploration and Mining on Earth to the Development of Off-World Natural Resources in the 21st Century, 1*, 1–24.
20. Chai, R., Savvaris, A., Tsourdos, A., Chai, S., & Xia, Y., (2019). A review of optimization techniques in spacecraft flight trajectory design. *Progress in Aerospace Sciences, 109*, 100543.

21. Chan, J., & DalBello, R., (2010). Data sharing to improve close approach monitoring and safety of flight. In: *Space Safety Regulations and Standards* (Vol. 1, pp. 133–136). Butterworth-Heinemann.
22. Cheng, G., Huang, Q., Xing, R., Lin, M., & Upadhyay, P. K., (2020). On the secrecy performance of integrated satellite-aerial-terrestrial networks. *International Journal of Satellite Communications and Networking, 38*(3), 314–327.
23. Chien, S., Sherwood, R., Tran, D., Cichy, B., Rabideau, G., Castano, R., & Boyer, D., (2005). Using autonomy flight software to improve science return on earth-observing one. *Journal of Aerospace Computing, Information, and Communication, 2*(4), 196–216.
24. Cox, C. B., & Coney, T. A., (1996). Advanced communications technology satellite adaptive rain fade compensation protocol performance. *NASA Technical Report, 1*, 1–20.
25. Dheepadharshani, S., Anandh, S., Bhavinaya, K. B., & Lavanya, R., (2019). Multivariate time-series classification for automated fault detection in satellite power systems. In: *2019 International Conference on Communication and Signal Processing (ICCSP)* (Vol. 1, pp. 0814–0817). IEEE.
26. Di Silvestre, M. L., Gallo, P., Guerrero, J. M., Musca, R., Sanseverino, E. R., Sciumè, G., & Zizzo, G., (2020). Blockchain for power systems: Current trends and future applications. *Renewable and Sustainable Energy Reviews, 119*, 109585.
27. Dunbabin, M., Corke, P., Winstanley, G. J., & Roberts, J. M., (2006). Off-world robotic excavation for large-scale habitat construction and resource extraction. In: *AAAI Spring Symposium: To Boldly Go Where No Human-Robot Team Has Gone Before* (Vol. 1, pp. 95–103).
28. Ejaz, M. Z., Khurshid, K., Abbas, Z., Aizaz, M. A., & Nawaz, A., (2018). A novel image encoding and communication technique of B/W images for IoT, robotics and drones using (15, 11) reed Solomon scheme. In: *2018 Advances in Science and Engineering Technology International Conferences (ASET)* (Vol. 1, pp. 1–6). IEEE.
29. Ellery, A., Kreisel, J., & Sommer, B., (2008). The case for robotic on-orbit servicing of spacecraft: Spacecraft reliability is a myth. *Acta Astronautica, 63*(5, 6), 632–648.
30. Emrick, R., Cruz, P., Carvalho, N. B., Gao, S., Quay, R., & Waltereit, P., (2014). The sky's the limit: Key technology and market trends in satellite communications. *IEEE Microwave Magazine, 15*(2), 65–78.

31. Esch, T., Heldens, W., Hirner, A., Keil, M., Marconcini, M., Roth, A., & Strano, E., (2017). Breaking new ground in mapping human settlements from space: the global urban footprint. *ISPRS Journal of Photogrammetry and Remote Sensing, 134*, 30–42.
32. Evans, B., Werner, M., Lutz, E., Bousquet, M., Corazza, G. E., Maral, G., & Rumeau, R., (2005). Integration of satellite and terrestrial systems in future multimedia communications. *IEEE Wireless Communications, 12*(5), 72–80.
33. Falkenmark, M., & Rockström, J., (2006). *The New Blue and Green Water Paradigm: Breaking New Ground for Water Resources Planning and Management, 1*, 1–29.
34. Farserotu, J., & Prasad, R., (2000). A survey of future broadband multimedia satellite systems, issues, and trends. *IEEE Communications Magazine, 38*(6), 128–133.
35. Fasano, G., Grassi, M., & Accardo, D., (2009). A stereo-vision-based system for autonomous navigation of an in-orbit servicing platform. In: *AIAA Infotech@ Aerospace Conference and AIAA Unmanned. Unlimited Conference* (Vol. 1, pp. 1934).
36. Fichman, R. G., Kohli, R., & Krishnan, R., (2011). Editorial overview: The role of information systems in healthcare: Current research and future trends. *Information Systems Research, 22*(3), 419–428.
37. Fleischauer, P. D., & Hilton, M. R., (1988). Assessment of the tribological requirements of advanced spacecraft mechanisms. *MRS Online Proceedings Library (OPL), 1*, 120–140.
38. Frey, K. E., & Smith, L. C., (2007). How well do we know northern land cover? Comparison of four global vegetation and wetland products with a new ground-truth database for West Siberia. *Global Biogeochemical Cycles, 21*(1), 1–33.
39. Godara, L. C., (1997). Applications of antenna arrays to mobile communications. I: Performance improvement, feasibility, and system considerations. *Proceedings of the IEEE, 85*(7), 1031–1060.
40. Gong, P., (2007). Wireless sensor network as a new ground remote sensing technology for environmental monitoring. *Journal of Remote Sensing-Beijing, 11*(4), 545.
41. Gritsuk, I., Gutarevych, Y., Mateichyk, V., & Volkov, V., (2016). *Improving the Processes of Preheating and Heating After the Vehicular Engine Start by Using Heating System with Phase-Transitional Thermal Accumulator* (No. 2016-01-0204, pp. 1–44). SAE Technical Paper.

42. Gubic, I., & Baloi, O., (2019). Implementing the new urban agenda in Rwanda: Nation-wide public space initiatives. *Urban Planning, 4*(2), 223–236.

43. Hagar, A., & Korolev, A., (2007). Quantum hypercomputation: Hype or computation? *Philosophy of Science, 74*(3), 347–363.

44. Hirzinger, G., Landzettel, K., Brunner, B., Fischer, M., Preusche, C., Reintsema, D., & Steinmetz, B. M., (2004). DLR's robotics technologies for on-orbit servicing. *Advanced Robotics, 18*(2), 139–174.

45. Hu, T., Huang, C., He, Y., Castillo-González, C., Gui, X., Wang, Y., & Zhou, X., (2019). βC1 protein encoded in Geminivirus satellite concertedly targets MKK2 and MPK4 to counter host defense. *PLoS Pathogens, 15*(4), e1007728.

46. Hu, X., Liu, S., Wang, Y., Xu, L., Zhang, Y., Wang, C., & Wang, W., (2019). Deep reinforcement learning-based beam Hopping algorithm in multibeam satellite systems. *IET Communications, 13*(16), 2485–2491.

47. Iida, T., & Pelton, J. N., (2003). The next thirty years. *Progress in Astronautics and Aeronautics, 202*, 169–190.

48. Iida, T., (2003). Cost consideration for future communications satellite. *Acta Astronautica, 53*(4–10), 805–810.

49. Iida, T., Ashford, E., & Pelton, J. N., (2003). *Satellite Communications in the 21st Century: Trends and Technologies* (Vol. 1, pp. 1–33). American Institute of Aeronautics and Astronautics.

50. Ippolito, L., Pelton, J. N., & Clarke, A. C., (2004). *Satellite Technology: The Evolution of Satellite Systems and Fixed Satellite Services* (Vol. 1, pp. 37–40). Lawrence Erlbaum Associates: New Jersey, USA.

51. Ji, X. Y., Li, Y. Z., Liu, G. Q., Wang, J., Xiang, S. H., Yang, X. N., & Bi, Y. Q., (2019). A brief review of ground and flight failures of Chinese spacecraft. *Progress in Aerospace Sciences, 107*, 19–29.

52. Jia, M., Zhang, X., Sun, J., Gu, X., & Guo, Q., (2020). Intelligent resource management for satellite and terrestrial spectrum shared networking toward B5G. *IEEE Wireless Communications, 27*(1), 54–61.

53. Kalita, H., Morad, S., Ravindran, A., & Thangavelautham, J., (2018). Path planning and navigation inside off-world lava tubes and caves. In: *2018 IEEE/ION Position, Location and Navigation Symposium (PLANS)* (Vol. 1, pp. 1311–1318).

54. Koren, Y., Gu, X., & Guo, W., (2018). Reconfigurable manufacturing systems: Principles, design, and future trends. *Frontiers of Mechanical Engineering, 13*(2), 121–136.
55. Levchenko, I., Bazaka, K., Belmonte, T., Keidar, M., & Xu, S., (2018). Advanced materials for next-generation spacecraft. *Advanced Materials, 30*(50), 1802201.
56. Liang, Y. W., Xu, S. D., & Tsai, C. L., (2007). Study of VSC reliable designs with application to spacecraft attitude stabilization. *IEEE Transactions on Control Systems Technology, 15*(2), 332–338.
57. Lindström, M., & Thornell, L. E., (2009). New multiple labeling method for improved satellite cell identification in human muscle: Application to a cohort of power-lifters and sedentary men. *Histochemistry and Cell Biology, 132*(2), 141–157.
58. Liolis, K., Geurtz, A., Sperber, R., Schulz, D., Watts, S., Poziopoulou, G., & Chuberre, N., (2019). Use cases and scenarios of 5G integrated satellite-terrestrial networks for enhanced mobile broadband: The SaT5G approach. *International Journal of Satellite Communications and Networking, 37*(2), 91–112.
59. Lopes, C. A., Prosser, S. L., Romio, L., Hirst, R. A., O'Callaghan, C., Woolf, A. S., & Fry, A. M., (2011). Centriolar satellites are assembly points for proteins implicated in human ciliopathies, including oral-facial-digital syndrome 1. *Journal of Cell Science, 124*(4), 600–612.
60. Lyon, D., (1998). The worldwide web of surveillance: The internet and off-world power-flows. *Information Communication and Society, 1*(1), 91–105.
61. Maclay, T., & Mcknight, D., (2020). Space environment management: Framing the objective and setting priorities for controlling orbital debris risk. *Journal of Space Safety Engineering, 1*, 1–33.
62. Makarova, I., Mukhametdinov, E., Mavrin, V., Shubenkova, K., & Garipov, R., (2018). Improvement of the vehicle's onboard diagnostic system by using the vibro-diagnostics method. In: *2018 International Conference on Diagnostics in Electrical Engineering (Diagnostika)* (Vol. 1, pp. 1–4). IEEE.
63. Malinowski, M., Leon, J. I., & Abu-Rub, H., (2017). Solar photovoltaic and thermal energy systems: Current technology and future trends. *Proceedings of the IEEE, 105*(11), 2132–2146.

64. Massari, M., & Zamaro, M., (2014). Application of SDRE technique to orbital and attitude control of spacecraft formation flying. *Acta Astronautica, 94*(1), 409–420.

65. Miao, Y., Cheng, Z., Li, W., Ma, H., Liu, X., & Cui, Z., (2016). Software defined integrated satellite-terrestrial network: A survey. In: *International Conference on Space Information Network* (Vol. 1, pp. 16–25). Springer, Singapore.

66. Navis, C., & Glynn, M. A., (2010). How new market categories emerge: Temporal dynamics of legitimacy, identity, and entrepreneurship in satellite radio, 1990–2005. *Administrative Science Quarterly, 55*(3), 439–471.

67. Norgren, P., Ludvigsen, M., Ingebretsen, T., & Hovstein, V. E., (2015). Tracking and remote monitoring of an autonomous underwater vehicle using an unmanned surface vehicle in the Trondheim fjord. In: *Oceans 2015-MTS/IEEE Washington* (Vol. 1, pp. 1–6). IEEE.

68. O'Dell, S. L., Atkins, C., Broadway, D. M., Elsner, R. F., Gaskin, J. A., Gubarev, M. V., & Zavlin, V. E., (2015). X-ray optics at Nasa Marshall space flight center. In: *EUV and X-Ray Optics: Synergy between Laboratory and Space IV* (Vol. 9510, p. 951003). International Society for Optics and Photonics.

69. Ogilvie, A., Allport, J., Hannah, M., & Lymer, J., (2008). Autonomous satellite servicing using the orbital express demonstration manipulator system. In: *Proc. of the 9th International Symposium on Artificial Intelligence, Robotics and Automation in Space (i-SAIRAS'08)* (pp. 25–29).

70. Ohata, K., Kobayashi, K., Nakahira, K., & Ueba, M., (2005). Broadband and scalable mobile satellite communication system for future access networks. *Acta Astronautica, 57*(2–8), 239–249.

71. Oteman, M., Wiering, M., & Helderman, J. K., (2014). The institutional space of community initiatives for renewable energy: A comparative case study of the Netherlands, Germany, and Denmark. *Energy, Sustainability and Society, 4*(1), 1–17.

72. Pecorella, T., Ronga, L. S., Chiti, F., Jayousi, S., & Franck, L., (2015). Emergency satellite communications: Research and standardization activities. *IEEE Communications Magazine, 53*(5), 170–177.

73. Pelton, J. N., (1998). *Telecommunications for the 21st Century* (Vol. 278, No. 4, pp. 80–85). Scientific American.

74. Pelton, J. N., (2005). *Future Trends in Satellite Communications: Markets and Services* (Vol. 1, pp. 1–43). Intl. Engineering Consortium.
75. Pelton, J. N., (2011). *Satellite Communications* (Vol. 1, pp. 1–32). Springer Science & Business Media.
76. Peng, Y., Dong, T., Gu, R., Guo, Q., Yin, J., Liu, Z., & Ji, Y., (2018). A review of dynamic resource allocation in integrated satellite and terrestrial networks. In: *2018 International Conference on Networking and Network Applications (NaNA)* (Vol. 1, pp. 127–132). IEEE.
77. Phadke, A. G., (1993). Synchronized phasor measurements in power systems. *IEEE Computer Applications in Power, 6*(2), 10–15.
78. Plaza, A., Plaza, J., & Vegas, H., (2010). Improving the performance of hyperspectral image and signal processing algorithms using parallel, distributed, and specialized hardware-based systems. *Journal of Signal Processing Systems, 61*(3), 293–315.
79. Rahmat-Samii, Y., & Densmore, A. C., (2014). Technology trends and challenges of antennas for satellite communication systems. *IEEE Transactions on Antennas and Propagation, 63*(4), 1191–1204.
80. Reintsema, D., Landzettel, K., & Hirzinger, G., (2007). DLR's advanced telerobotic concepts and experiments for on-orbit servicing. In: *Advances in Telerobotics* (Vol. 1, pp. 323–345). Springer, Berlin, Heidelberg.
81. Reynolds, W. D., Jeon, S. K., Banik, J. A., & Murphey, T. W., (2013). Advanced folding approaches for deployable spacecraft payloads. In: *International Design Engineering Technical Conferences and Computers and Information in Engineering Conference* (Vol. 55942, p. V06BT07A043). American Society of Mechanical Engineers.
82. Rovetto, R. J., (2016). An ontological architecture for orbital debris data. *Earth Science Informatics, 9*(1), 67–82.
83. Ruan, Y., Li, Y., Wang, C. X., & Zhang, R., (2017). Energy-efficient adaptive transmissions in integrated satellite-terrestrial networks with SER constraints. *IEEE Transactions on Wireless Communications, 17*(1), 210–222.
84. Ruiz-Garcia, L., Lunadei, L., Barreiro, P., & Robla, I., (2009). A review of wireless sensor technologies and applications in agriculture and food industry: State of the art and current trends. *Sensors, 9*(6), 4728–4750.
85. Sachdev, D. K., (2004). *Business Strategies for Satellite Systems* (Vol. 1, pp. 1–24). Artech House.

86. Sarda, K., Eagleson, S., Caillibot, E., Grant, C., Kekez, D., Pranajaya, F., & Zee, R. E., (2006). Canadian advanced nanospace experiment 2: Scientific and technological innovation on a three-kilogram satellite. *Acta Astronautica, 59*(1–5), 236–245.

87. Sarda, K., Grant, C., Eagleson, S., Kekez, D. D., & Zee, R. E., (2010). Canadian advanced nanospace experiment 2 orbit operations: Two years of pushing the nanosatellite performance envelope. In: *ESA Small Satellites, Services and Systems Symposium* (Vol. 20, pp. 1–33).

88. Saunders, K., Norman, A., Gucciardo, S., & Stanley, J., (2004). The DNA β satellite component associated with ageratum yellow vein disease encodes an essential pathogenicity protein (βC1). *Virology, 324*(1), 37–47.

89. Sheard, B. S., Heinzel, G., Danzmann, K., Shaddock, D. A., Klipstein, W. M., & Folkner, W. M., (2012). Intersatellite laser ranging instrument for the GRACE follow-on mission. *Journal of Geodesy, 86*(12), 1083–1095.

90. Sherwood, R., Chien, S., Tran, D., Cichy, B., Castano, R., Davies, A., & Rabideau, G., (2003). *Next-Generation Autonomous Operations on a Current Generation Satellite, 1,* 1–32.

91. Skinner, M. A., (2017). Orbital debris: What are the best near-term actions to take? A view from the field. *Journal of Space Safety Engineering, 4*(2), 105–111.

92. Smyth, T. N., Etherton, J., & Best, M. L., (2010). MOSES: Exploring new ground in media and post-conflict reconciliation. In: *Proceedings of the SIGCHI Conference on Human Factors in Computing Systems* (Vol. 1, pp. 1059–1068).

93. Subudhi, B., & Pradhan, R., (2012). A comparative study on maximum power point tracking techniques for photovoltaic power systems. *IEEE Transactions on Sustainable Energy, 4*(1), 89–98.

94. Taleb, T., Kato, N., & Nemoto, Y., (2005). Recent trends in IP/NGEO satellite communication systems: Transport, routing, and mobility management concerns. *IEEE Wireless Communications, 12*(5), 63–69.

95. Toyoshima, M., (2005). Trends in satellite communications and the role of optical free-space communications. *Journal of Optical Networking, 4*(6), 300–311.

96. Vasavada, Y., Gopal, R., Ravishankar, C., Zakaria, G., & BenAmmar, N., (2016). Architectures for next generation high throughput satellite

systems. *International Journal of Satellite Communications and Networking, 34*(4), 523–546.

97. Venkatesan, R., Arul, M. M., Ramesh, K., Ramasundaram, S., Sundar, R., & Atmanand, M. A., (2013). Satellite communication systems for ocean observational platforms: Societal importance and challenges. *Journal of Ocean Technology, 8*(3), 1–32.
98. Voigt, S., Giulio-Tonolo, F., Lyons, J., Kučera, J., Jones, B., Schneiderhan, T., & Guha-Sapir, D., (2016). Global trends in satellite-based emergency mapping. *Science, 353*(6296), 247–252.
99. Wasser, S. P., (2011). Current findings, future trends, and unsolved problems in studies of medicinal mushrooms. *Applied Microbiology and Biotechnology, 89*(5), 1323–1332.
100. Winn, W., (2002). Research into practice: Current trends in educational technology research: The study of learning environments. *Educational Psychology Review, 14*(3), 331–351.
101. Woellert, K., Ehrenfreund, P., Ricco, A. J., & Hertzfeld, H., (2011). CubeSats: Cost-effective science and technology platforms for emerging and developing nations. *Advances in Space Research, 47*(4), 663–684.
102. Yang, Y., Su, P., & Tang, Y., (2018). Stimuli-responsive lanthanide-based smart luminescent materials for optical encoding and bio-applications. *Chem. Nano Mat., 4*(11), 1097–1120.
103. Yin, S., Xiao, B., Ding, S. X., & Zhou, D., (2016). A review on recent development of spacecraft attitude fault-tolerant control system. *IEEE Transactions on Industrial Electronics, 63*(5), 3311–3320.
104. You, T., Shi, W., Wang, B., Wu, W., & Li, X., (2015). A new ground-coupled heat pump system integrated with a multi-mode air-source heat compensator to eliminate thermal imbalance in cold regions. *Energy and Buildings, 107*, 103–112.
105. Yu, G., Vladimirova, T., & Sweeting, M. N., (2009). Image compression systems on board satellites. *Acta Astronautica, 64*(9, 10), 988–1005.

INDEX

Symbols

3D 208

A

Aacquisition 189
Acceleration zones 145
Accelerator grids 141
AceS (Asian cellular system) 62
Adequate maturity 10
Aerospace Corporation 136
Air resistance 46
Algorithm 115
Allocations 209
Altimetry technology 161
Ambient (space) environment 143
Analogous 138, 198
Anode 145
Antenna fabrication 193
Antennas 16, 49, 107
Antenna systems 207
Approximation 156
Argon (Ar) 146
Asteroids 45
Astronomy 46
Astrophysical 3
Atmospheric parameters 4
Atmospheric situations 197
Attenuation 207
Automobile 115
Autonomous operations 208

B

Beams 84
Belt telephone laboratory 116
Bi-propellant 131

C

Cameras 25
Cannibalization 74
Capability 168
Capacity 199

Carbon cycle 33
Carbon fiber 176
Cartography 24
Cathode 142, 144
CGP system 128
Chamber pressure (Pc) 135
Chamber temperature (Tc) 129
Characteristic velocity (C) 129, 136
Chronic gains 188
Circular orbits 78
Climate forecasts 161
Cloud formation 23
Coders 197
Cold gas propulsion (CGP) 128
Collaborative data 105
Collisions 79
Combustion chamber temperature (Tc) 132
Combustion process 135, 136
Commercial-off-the-shelf (COTS) 126
Communicate 164
Communication 2, 126
Communications links 101
Communication systems 72, 119
Communicators 84
Community 126
Comstar satellite 19
Congestion 115
Constellation 55, 100, 167
Consumer terminals 120
Crop production 24
Cryosphere 33
Cube-Satellites 126
CubeSats 166
Customers 89
Customization 133
Cyclic nature 118

D

Data communications 63
Decoders 197
Decompose 131
Decrease 156
Degradation 146
Delay-sensitive 100
Delta-v (Δv) 127
Density 139, 144
Departments 156
Depletion 88
Deployment system 166
Designers 86
De-stabilizing effect 130
Destination 83
Diameters 191
Digital data 13
Digital processors 188
Dirigible 107
Dirigible spot 107
Discharge plasma 144
Discrete orbits 46
Domestic 116
Dramatic gains 188
Dynamic data 105
Dynamic environments 85

E

Earth's rotation period 114
Economically viable 74
Education 126
Efflux 142
EGC (enhanced group calls) 63
EHIS (energetic heavy ion sensor) 32
Electric 144
Electrical discharge (E˜) 145
Electric field 144
Electric propulsion systems 126
Electric solid propellant (ESP) 137
Electrochromic windows 143
Electrode 143
Electromagnetic spectrum 197

Electron-bombardment 143
Electron energy 146
Electronics 2
Electrons 141, 145, 159
Electrostatic grid 141
Elliptical 14, 78
Empower 157
Encrypted data 105
Endeavors 207
Energy 87
Energy sources 86
Energy systems 202
Enterprise 188
Enthusiasts 2
Epoxy 137
Equator 114
Equatorial plane 117, 119
Examination 156
Execution 107
Exit-to-chamber-pressure (Pe/Pc) 129
Exit-to-chamber-pressure-ratio (Pe/Pc) 132
Expendable 30
External cathode 144
External forces 118

F

Fading 119
FCC (Federal Communications Commission) 48
Femto 156
Femto-satellites 160
Fiber transmissions 198
Finance 157
Flashings 119
Flexibility 44
Flexible 47
Flow rate 135
Forecasters 190
Forestry 24
Formation 164
Frequency ranges 191
Fuel 202

G

Gaseous propellants 129
Gathering 76
GCR (galactic cosmic rays) 159
Generation 20
Generator 144
Geographically 191
GEO satellite systems 44
Geostationary 3, 156
Geostationary meteorological satellite (GMS) 22
Geostationary orbit (GEO) 6
Geostationary satellites 55, 63, 114
Geostationary transfer orbit (GTO) 30
Geosynchronous 48, 117
Geosynchronous earth orbit satellites (GEO) 48
Geosynchronous network 114
Geosynchronous orbit 48
Geosynchronous satellite 50
Globalstar space 61
Governmental organizations 157
Graduate students 169
Gravity 33, 45, 118

H

Hardware 83
Heavier 156
HEO (highly elliptical orbit) 56
High-resolution 25
Hobbyists 2
Hurricanes 23
Hydrazine 131
Hydrazine thrusters 133
Hyperspectral sensors 32

I

Ignitability 133
Implementation 6
Impulse 129
Impulse propulsion system 140
Inclination 57, 131
Increasing 108
Infrastructure 61
Inherent 44
Initial (mi) 127
Initial phase 34
Inner wall 146
Innovative 203
Instantaneous 136
Integration 209
Intelsat-1 14
Intensive frequency 191
Interactive navigation 208
Interactivity 104
Intercontinental communication services 12
Intercontinental satellite 2
Internal charging 176
Internet engineering task force (IETF) 206
Interplanetary 47
Inter-satellite 81
Inter-satellite space 84
Intersystem interference 209
InteSractive 103
Investors 86
Ion beam current (Ii) 142
Ion bombardment 143
Ionizing 158, 159
ISR platform 107

K

Kinetic energy 127

L

Lagrange points (L-points) 47
LEO constellations 120
LEO (low-Earth orbit) 159
LEOS system 86, 87
Line of sight (LOS) 76
Liquid propulsion (LP) system 131
Locations 48
Low earth orbit satellites (LEO) 48
Lower efficiency 144
Low power consumption 139
Low-toxicity 133
LP systems 131

M

Magnetic attraction 194
Magnetic circuitry 144
Magnetic fields 144
Magnetic shielding (MS) 146
Manufacturing 172
Marginally 44
Maritime clients 53
Mass 135
Mathematical relations 129
Measurement 176
Medium earth orbit (MEO) 51, 88, 100
Medium earth orbit satellites (MEO) 48
Melting 138
Message 84
Metamorphic movement 207
Meteorological satellites 21
Micro 156
Micromechanical alterations 176
Micro-propulsion system 130, 134
Microsatellite 162
Micro-thrusters 136
Microwave 116

Microwave power 142
Microwave thrusters 141
Miniaturization 205
Minisatellite 160
Missile defense alarm system (MIDAS) 10
Mitigation 158
Mixture ratio 133
Momentum 127
Monitoring 73
Mono-propellant 131
MSS (multispectral scanners) 25
Multi-beam 203
Multiple satellites 46
Multiplexing 197
Multispectral imagers 32
Multispectral imagery 107

N

Nano 156
Nanosatellite 164
Neutralize 141
Neutralizer 143
Neutralizer cathode 141
Neutralizing electrons 142
Nitrous oxide 131
Nominal magnetic field 146
Non-electric propulsion systems 126
Non-electric systems 126
Non-geosynchronous 100
Non-geosynchronous orbit 12
Non-ionized propellant 142
Novel material 176
Nozzle 135
Nozzles form 127
Nuclear energy 201

O

Onboard computers (OBCs) 208
Onboard switching 207
Optimization 86
Orbit 44, 160
Orbital configurations 189
Orbit Satellites 47
Orifice plate 143
Oxidizer 131, 135, 137

P

Parabolic reflector 194
Parameters 8
Particular system 73
Percentage 75
Photovoltaic (PV) 189
Picosatellite 168
Polarization 16, 102
Polar orbit 47
Political independence 84
Pollution monitoring 24
Portable 205
Power generation 79
Power supplies 205
Probability 139
Procedure 108
Process 128
Production 156, 157
Productivity 105
Propagation 79
Propellants 131
Propellant storage 131
Propulsion 139
Propulsion system 28, 126
Propulsive maneuvers 131
Protocols 115
Publication 74

Q

Qualification 146
Quantum 202

R

Radiation dose 158

Radioactive materials 201
Radiofrequency ion thrusters 141
Radio frequency (RF) energy 189
Radio satellites 115
Radius 3
Rain attenuation 196
Rainbow solar cells 200
Rapidly 206
Reallocation 189
Receivers 33
Reduction 146
Regarding dark 88
Regeneration 32
Regulatory 209
Regulatory change 189
Relaying 84
Reliability 191
Resolutions 158
Responders 101
Retransmission 105
Reusable 30
Re-utilization 16
Revenue losses 88
Rocket 168
Rocket, satellites 88
RSS (remote sensing satellites) 23

S

SAR (synthetic aperture radar) 25
Satellite 2
Satellite communications 202
Satellite communication system 44
Satellites 157
Satellite system 18
Scientific data 156
SEISS (space environment in-situ suite) 32
Self-reliant 26
Sensitivity 102
Server 104
Smaller satellites 157
Smartphones 168
Software 83
Solar cells 199
Solar panels 86
Solar radiation 199
Solid propellant 135
Solid Propellant Micro-thrusters (SPMs) 136
Solid rocket propulsion (SRP) 135
Solid-state Ionics (SSIs) 143
Spacecraft 130, 132, 176
Spacecraft instruments 142
Spacecraft materials 175
Space exploration 2
Space radiation 159
Spectra 162, 198
Spectrum 23, 44, 197
Spot beams 107
Spying 2
SSO (Sun-synchronous orbit) 58
Stabilization 8, 50, 84
Station-keeping 118
Steering 119
Streams 195
Subscription 86
Successive firing 136
Sulfur Hexafluoride (SF6) 133
Summarizing 136
Sun-synchronous orbit (SSO) 47
Superconductors 143
Superiority 26
Surrey Satellite Technology Ltd. (SSTL) 130
Switching 80
Synchronous 49, 58

T

Technocrats 2
Technology 54
Telecommunication 46, 72

Telecommunication services 89
Telecommunications systems 100, 188
Telephony 2
Telex 63
Telstar 52
Temperature 138, 139, 176
Temperature coefficient (a) 135
Terminal 107
Terrestrial cable 197
Terrestrial data 208
Terrestrial fiber 103
Terrestrial structures 44
Terrestrial wireless 198
Thermal losses 138
Thickness 175
Thruster 142
Thrust regulation 136
TM (thematic mappers) 25
Topography 115
Toxicology 132
Tracking 107, 207
Tracking antenna 107
Transfer 107, 108
Transfer orbits 59
Transferring 108
Transmission 61, 86, 192
Transmission beams 197
Transmissions 13
Transmission systems 206
Transmitter 116
Transponders 16, 53
Transportation 129
Tremendous 200
Trinity 115
Tropospheric proliferation 119
Tropospheric reduction 120

U

Ultra-capacitor 172
Unnatural quality 103

V

Validation 126
Vegetation 115
Vehicles 160
Velocities 85
Velocity 117
Viability 75
Videoconferences 104
Virtual reflector 194
Vital role 86
Voltage 142

W

Wall erosion 146
Wavelengths 193
Wearing 144
Westar satellite 19
Wind resistance 190
Wireless interconnection 188

X

Xenon 146

Y

Yttrium oxide 176